Palgrave Studies in Cyberpsychology

Series Editor
Jens Binder
Nottingham Trent University
Nottingham, UK

Palgrave Studies in Cyberpsychology aims to foster and to chart the scope of research driven by a psychological understanding of the effects of the 'new technology' that is shaping our world after the digital revolution. The series takes an inclusive approach and considers all aspects of human behaviours and experiential states in relation to digital technologies, to the Internet, and to virtual environments. As such, Cyberpsychology reaches out to several neighbouring disciplines, from Human-Computer Interaction to Media and Communication Studies. A core question underpinning the series concerns the actual psychological novelty of new technology. To what extent do we need to expand conventional theories and models to account for cyberpsychological phenomena? At which points is the ubiquitous digitisation of our everyday lives shifting the focus of research questions and research needs? Where do we see implications for our psychological functioning that are likely to outlast shortlived fashions in technology use?

Davide Maria Marchioro
Aimée Argüero Fonseca
Fabio Benatti • Marco Zuin

Virtual Reality: Unlocking Emotions and Cognitive Marvels

Methodology, Tools and Applications

Davide Maria Marchioro
Department of Psychology
Istituto Universitario Salesiano
di Venezia
Venezia, Italy

Fabio Benatti
Department of Psychology
Istituto Universitario Salesiano
di Venezia
Venezia, Italy

Aimée Argüero Fonseca
Psychology Department
Autonomous University of Nayarit
Tepic, Mexico

Marco Zuin
Department of Psychology
Istituto Universitario Salesiano
di Venezia
Venezia, Italy

ISSN 2946-2754   ISSN 2946-2762   (electronic)
Palgrave Studies in Cyberpsychology
ISBN 978-3-031-68195-0   ISBN 978-3-031-68196-7   (eBook)
https://doi.org/10.1007/978-3-031-68196-7

This Palgrave Macmillan imprint is published by the registered company Springer Nature Switzerland AG.
The registered company address is: Gewerbestrasse 11, 6330 Cham, Switzerland

If disposing of this product, please recycle the paper.

PREFACE

Technological advancement has always represented a frontier for the expansion of human knowledge, and virtual reality (VR) is no exception. In this volume, we will explore the intersection between psychology and VR, revealing the extraordinary potential this technology holds in the field of psychological sciences. VR, once relegated to the realm of science fiction, now stands as a tangible and accessible tool capable of transforming the way we understand and influence human behavior, cognition, and emotions.

My encounter with VR as a therapeutic tool happened almost by chance. Initially skeptical of its effectiveness, I was quickly convinced by the changes I observed in some patients treated with the aid of this technology. One of the first cases that particularly struck me involved a patient suffering from agoraphobia, whom I referred to a colleague specializing in cognitive-behavioral therapy (CBT). This colleague had recently incorporated the use of virtual reality into his therapeutic approach. Given the severity of the patient's condition, traditional CBT methods would have required months of gradual exposure therapy to confront her fears. However, with the adoption of virtual reality, it was possible to accelerate and intensify the therapeutic process, while still remaining within a controlled and protected environment.

I still vividly remember the video documenting her first "virtual session," set in a crowded park. The patient's initial anxiety was evident, but it seemed as if the ability to "exit" the virtual environment at any moment infused her with a sense of trust and control that she had rarely experienced in her daily life. Over the course of several sessions, her confidence grew exponentially. In this and other cases, the use of VR was not merely limited to replicating real-world scenarios, but also provided a safe stage on which the patient could easily confront her most challenging fears, ultimately rewriting her personal narrative characterized by anxiety and fear.

VR also affects the therapist, enabling direct observation of patients' reactions and the adaptation of sessions in real time to maximize treatment efficacy. However, the adoption of this technology is not without challenges. The advancement of VR and its recent introduction into the field of psychology, and more specifically, psychotherapy, represent not just a technical evolution but also a true paradigm shift that prompts deep ethical reflections. The issue of the authenticity of human experiences within digitally recreated worlds arises: the ability of VR to "deceive the mind," creating immersive environments capable of eliciting genuine emotional responses, inevitably raises several questions about the concept of reality and what it truly means to have "real" experiences.

The very concept of "immersion" brings to mind Plato's allegory of the cave, where individuals confuse the shadows cast on the wall with reality itself. In a sense, VR creates a sort of "modern cave," in which shadows are replaced by digital simulations. However, unlike Plato's allegory, the aim of VR should be to illuminate and expand our understanding of psychological and therapeutic reality, rather than deceive us.

The ability of VR to simulate complex environments and elicit authentic emotional responses thus invites us to reflect more deeply on the nature of our perception and cognitive experiences. VR offers an unprecedented laboratory for observing how the human brain interacts with realities that, although artificial, provoke reactions quite similar to those that would occur in a non-virtual context. This leads us to consider the mind not only as a biological entity but also as a dynamic and flexible construct, capable of adapting to and being influenced by virtual environments.

With the power to create highly immersive and personalized experiences, VR compels us to rethink the very concept of therapy and psychological well-being: while the use of virtual environments for mental health opens new avenues for treating certain emotional and cognitive disorders, it also forces us to reflect on what it truly means to "heal" in an era dominated by augmented realities and digital worlds. Therefore, it is essential to proceed with caution, ensuring that we do not reduce the complexity of human experience to mere simulations. Each session should serve as a reminder that, while it is possible to embrace these extraordinary opportunities with enthusiasm, we must also remain anchored to the fundamental principles of empathy and human integrity in patient care.

The responsibility of psychologists and researchers is thus to navigate these new territories with care and awareness, ensuring that VR applications are not only effective but also ethically sustainable. This book, therefore, is not merely a technical exploration of VR applications in psychology but also an invitation to reflect on the deeper implications of these emerging technologies. We invite readers to join us on this fascinating journey to discover the practical potentials of VR, without forgetting the philosophical and ethical issues that arise in this exciting new field of study.

Venice, Italy   Davide Maria Marchioro

CONTENTS

LIST OF FIGURES

LIST OF BOXES

LIST OF TABLES

CHAPTER 1

Introduction

Abstract This chapter explores the transformative potential of Virtual Reality (VR) in the realm of psychology, emphasizing its applicative reach from therapeutic settings to educational environments. VR technology not only augments traditional psychological practices but also redefines them, offering immersive experiences that enhance patient engagement and emotional presence. These capabilities make VR an invaluable tool in treating anxiety disorders, phobias, and PTSD, utilizing controlled, replicable virtual environments that simulate real-world challenges without the associated risks. Additionally, VR's application extends to educational psychology, where it improves learning outcomes through simulated, multisensory educational environments. The chapter discusses both the technological aspects of VR, such as immersion and presence, and its psychological impacts, including enhanced cognitive processing and emotional response. It concludes by examining the prospects of VR in psychological practice, emphasizing ongoing research and ethical considerations.

Keywords Virtual reality • Cyberpsychology • Clinical applications • Virtual environments • Sense of presence

Premise The rapid advancement of digital technologies, combined with the pervasive adoption of mobile devices, has made the digital realm an integral part of our lives to the extent that younger generations perceive digital technology as an essential component of their daily existence, akin to a natural extension of their cognitive and sensory capabilities. This shift has profoundly transformed the way we establish social relationships, communicate, and behave, prompting a reevaluation of our understanding of identity, presence, and human interaction (Turkle, 2011).

In this context, virtual reality (VR) emerges as an advanced form of human-computer interface, allowing users to interact with and immerse themselves in computer-generated simulated environments. VR offers an experience of sensory and perceptual delocalization, in which the user is completely immersed in a virtual environment. This technology has found significant applications in psychotherapeutic practice, enabling a tangible and dynamic representation of the patient's inner world. However, it is important to emphasize that VR does not replace traditional processes of constructing the patient's imagination but rather enhances them, serving as an integrative therapeutic tool useful in specific situations. When a patient has difficulty vividly imagining and recalling emotions, VR can be employed as a complementary tool to facilitate these processes (Freeman et al., 2017; Maples-Keller et al., 2017). As Riva (2005) suggests, VR can indeed be conceived as an enhancement of traditional intervention rather than a mere substitute for its fundamental components (Wiederhold & Bouchard, 2014).

The ability of VR to create immersive experiences has enabled psychologists to study cognitive and emotional responses in ways previously unattainable with traditional methods. For instance, VR has been effectively utilized in the treatment of anxiety disorders, providing safe environments for exposure therapy that are both controllable and repeatable, thereby reducing the logistical and ethical constraints inherent in real-world settings (Botella et al., 2015). Moreover, research suggests that the immersive nature of VR can lead to higher levels of emotional engagement and cognitive presence, which are crucial in therapeutic settings and psychological experiments (Turner & Casey, 2014). This potential reflects a phenomenological view of experience, where the lived experience of the subject is central to understanding cognitive and emotional dynamics.

In addition to clinical applications, VR is also proving to be a revolutionary tool in the field of educational psychology. As demonstrated by the study of Parsons and Rizzo (2008), VR not only enhances learning and retention through the simulation of multisensory educational environments but can also be used to assess and improve attentional processing capabilities in educational contexts. Using virtual environments, such as a simulated city and a driving simulation, detailed neuropsychological tests can be conducted to measure attention and the recall of specific targets.

Recent studies further support these findings. For example, research conducted by Makransky and Lilleholt (2018) demonstrated that the use of VR in educational settings significantly increases student engagement and motivation, thereby improving learning outcomes (Makransky & Lilleholt, 2018). Similarly, Lin et al. (2024) highlighted that the integration of VR into science lessons led to an increase in students' conceptual understanding and problem-solving abilities (Lin et al., 2024). The studies conducted thus far have essentially demonstrated that by increasing the complexity and intensity of stimuli, students' attentional performance can be effectively manipulated and evaluated. This shows how VR can create complex and immersive educational scenarios that not only enhance understanding and memory of the information presented but also allow students to experience learning situations firsthand, which is crucial for cognitive development and information management.

These VR-created scenarios, which simulate the complexity and dynamics of the real world, allow teachers and researchers to explore new paradigms of teaching and learning. This potential paves the way for an education that is no longer limited by physical context but can be enriched by virtual experiences that stimulate students' curiosity and engagement. VR, therefore, is not just a technological tool but represents a new epistemological frontier for education, enabling the exploration and understanding of knowledge in innovative ways.

This first chapter, general and introductory in nature, aims to explore how VR is rewriting methodologies in various fields of psychology by examining its potential. We will specifically analyze its role within traditional psychological practices, assessing how VR can enhance both therapeutic and research applications. Additionally, we will explore the effect of VR on emotions and cognitive functions, investigating how immersive experiences modulate fundamental psychological processes. Various aspects will be addressed, from psychodiagnostic assessment to the use of VR for cognitive engagement and the management of cognitive fatigue. The impacts of VR on children's motivation and on the treatment of grief and emotional loss through mindfulness-based interventions will also be discussed. Finally, the volume will conclude by examining the current limitations and prospects of VR in psychology, with particular attention to ethical considerations.

The objective is to provide an in-depth and updated view of the use of VR in psychology, contributing to the development of new therapeutic and educational strategies and stimulating further research in this emerging field. VR is not merely a technology but a new dimension of our existence, inviting us to reflect on what it means to be human in an increasingly digital world.

1.1   The Potential of Virtual Reality in Psychology and Its Link to Cyberpsychology

In contemporary times, the intersection between psychology and VR has catalyzed the development of innovative methodologies for the analysis and treatment of psychological disorders. The adoption of VR in the context of psychological practices, a central theme in cyberpsychology, has deepened our understanding of human behaviors within digital environments. Unlike traditional psychology, which focuses on the analog reality, cyberpsychology is dedicated to studying human interaction with digital realities, exploring the influence of technologies on the human psyche. It is a discipline particularly concerned with how technologies can be optimized to enhance or compensate for individual capabilities and vulnerabilities, significantly impacting people's daily lives. VR thus emerges as a valuable tool in cyberpsychology, primarily because it offers the possibility of creating highly controlled and detailed virtual environments that faithfully mimic realistic scenarios, providing a unique platform to safely and reproducibly simulate and study complex everyday situations (Rubio-Tamayo et al., 2017; Freeman et al., 2017).

Recent studies have expanded this view, suggesting that VR not only accurately reproduces real scenarios but also facilitates psychological intervention in environments that would otherwise be inaccessible or less controllable, thereby improving the effectiveness of therapies for disorders such as anxiety, PTSD, and phobias (Botella et al., 2017b; Lindner et al., 2017). These studies underscore the transformative role of VR in shaping new frontiers in psychological therapy and behavioral investigation.

Box 1.1 Virtual Reality: Key Concepts and Recent Developments
The term virtual reality (VR) was coined in 1989 by computer visionary Jaron Lanier, founder of the first virtual reality industry (Krueger, 1991). Initially, this technology was conceived as a combination of hardware and software capable of creating interactive three-dimensional environments that simulate everyday life experiences (Riva, 2006; Schultheis & Rizzo, 2001). There are many definitions of virtual reality, but the most widely accepted by the scientific community considers it as a system of computer devices capable of generating a new type of human-computer interaction (Steuer, 1992; Ellis, 1994).

Two key concepts are associated with the use and implementation of virtual reality: the **degree of immersion** and **presence** (Riva et al., 2003).

From a technological perspective, *immersiveness* defines the extent (from "non-immersive" to "fully immersive") to which the user is isolated from the real world while interacting with virtual environments. Fully immersive systems involve complete sensory absorption of users into the computer-generated three-dimensional world, typically through virtual headsets and position sensors (trackers). These are advanced body tracking systems that allow synchronization between user movements and corresponding changes in virtual environments in real time.

From a psychological perspective, the added value of virtual reality compared to other experiences is the *sense of presence*, that is, the sensation of "being really there" within the digitally created environment that replaces real perceptions (Riva, 2018).

Recent studies continue to explore and confirm these concepts. For instance, Slater and Sanchez-Vives (2016) discuss the importance of presence in VR, defining it as a critical component for the effectiveness of VR applications in fields such as psychological therapy and professional training. Similarly, a study by Cummings and Bailenson (2015) highlights how the degree of technological immersion directly affects the sense of presence, with significant implications for the design of user interfaces and virtual experiences.

1.1.1 The Sense of Presence in Virtual Reality

The "sense of presence" in VR represents the subjective perception of being physically present in a virtual environment. This concept, central to the effectiveness of VR applications, is defined as the impression of "being there" in the virtual space, despite the awareness of being physically located elsewhere. According to Slater (2009), presence is a psychological construct that emerges when immersive technologies fully engage the user's senses, tricking the mind into perceiving the virtual world as real.

The sense of presence is influenced by various factors, including the quality of image and sound, movement latency, and the interactivity of the virtual environment. Higher visual resolution and accurate spatial audio certainly contribute to making the experience more realistic and immersive, but what further strengthens the sense of presence is the ability to interact in real time with the virtual environment, for example, through hand or body movement.

The sense of presence is experienced when the user ceases to perceive the existence of the technology that generated the virtual environment (such as the VR headset), and simultaneously, the environment responds as if the technology were absent. Pavel Zahoric and Rick Jenison (1998) argue that the sense of presence is connected to the ability to effectively perform actions in the environment, whether physical or digital (Zahoric & Jenison, 1998). Additionally, as highlighted by Lakshmi Sastry and David Boyd (1998), a virtual environment exhibits a high level of presence when the user can *intuitively* navigate, select, move, and manipulate objects (Sastry & Boyd, 1998).

But what is meant by intuition? From a psychological perspective, intuition represents the capacity for simulation, which is the mental ability to anticipate, before undertaking a specific action or movement, all the necessary steps to adequately complete the intended task; this includes activities such as riding a bicycle, driving a car, or kicking a ball. Humans can perform these actions intuitively because the mind uses learned motor patterns to predict potential situations and adapt accordingly. In a video game, if one can control and act through the avatar based on intuitive thought, then the mind "embodies" itself in the virtual space surrounding the avatar, and from that exact moment, one is effectively inside the video game. Furthermore, according to Riva and Gaggioli (2019), for the mental simulation system, the perception of being "present" in a space is determined by the ability to act (Riva & Gaggioli, 2019): presence manifests where one can operate intuitively (Riva, 2012).

Using the sense of presence induced by virtual reality makes it easier to develop new, realistic, and credible informative experiences about the surrounding world and to demonstrate to the individual that what they believe to be true is actually a construct of their mind (Riva et al., 2016). The action possibilities offered by VR, which provide the subject with the perception of being present at a given moment and context, allow them to experience, both cognitively and emotionally, the crucial transition from the role of passive receiver and observer of an experience to that of an active protagonist.

A significant contribution to the understanding of the concept of "presence," as well as the first attempt to measure its construct, comes from the work of Witmer and Singer (1998), who developed the *Presence Questionnaire* (PQ) and the *Immersive Tendencies Questionnaire* (ITQ) to measure the sense of presence in virtual environments. While the PQ assesses the degree of presence experienced by individuals in a virtual environment and the influence of various factors on the intensity of this experience, the ITQ measures the individual tendency to experience presence.

Witmer and Singer (1998) define "presence" as the subjective experience of being in one place or environment, even when physically situated in another. They consider presence to be a normal awareness phenomenon that requires direct attention and is based on the interaction between sensory stimulation, environmental factors that promote engagement, and the ability to immerse oneself. The authors identified several factors that contribute to presence, grouping them into main categories: control factors, sensory factors, distraction factors, and realism factors (Witmer & Singer, 1998).

The experiments conducted by Witmer and Singer demonstrated that the PQ and ITQ are internally consistent and highly reliable measures. They found a weak but consistent positive relationship between presence and performance in virtual tasks. Additionally, they discovered that individual tendencies, as measured by the ITQ, predict presence as measured by the PQ, and that simulator sickness symptoms are negatively correlated with presence.

A critical aspect of the study to consider is the identification of the necessary conditions for experiencing presence. Witmer and Singer emphasize that engagement is a psychological state that arises from focusing energy and attention on a coherent set of meaningful stimuli or activities: the greater the attention directed toward the stimuli of the virtual environment, the greater the engagement and, consequently, the sense of presence.

Immersion, defined as the psychological state characterized by the perception of being enveloped and included in an environment that provides a continuous flow of stimuli and experiences, is another determining factor. A virtual environment that effectively isolates users from the physical environment, providing natural interaction modes and perception of movement, increases the degree of immersion and, therefore, presence.

Witmer and Singer's (1998) study has significantly contributed to the understanding of the sense of presence in virtual reality, highlighting the importance of engagement and immersion as key components and paving the way for subsequent research. Their measurements and analyses have indeed provided a solid foundation for further investigations into optimizing virtual environments to enhance user experience and the therapeutic and educational applications of VR.

Another example of research that assessed the sense of presence is the study conducted by Usoh et al. (2000), which compared the effectiveness of two presence questionnaires: the previously mentioned PQ by Witmer and Singer (Witmer & Singer, 1998) and the "Slater-Usoh-Steed Questionnaire" (SUS). The study involved twenty participants divided into two groups: one group searched for an object in a real environment, while the other performed the same task in a virtual environment modeled on the same physical space. Participants completed both questionnaires immediately after the experience. The results showed that the WS was not significantly able to distinguish between real and virtual experiences, whereas the SUS showed a slight superiority in discriminating between the two environments, although not consistently for all questions. This suggests that while presence questionnaires can be useful for evaluating immersion in a single type of environment, their utility for comparisons between real and virtual environments remains questionable (Usoh et al., 2000).

The study also highlighted that participants reported a stronger sense of presence when the virtual environment was interactive and responded realistically to their actions, in line with the findings of Witmer and Singer (1998), which demonstrated that interaction and focused attention were crucial components of the sense of presence. Emotional and cognitive engagement of the user thus emerges as a determining factor for the perception of presence, making virtual experiences more immersive and realistic.

PQ has also been employed in numerous other studies to evaluate the effectiveness of different VR technologies and configurations, providing a standardized method to compare levels of presence across various virtual experiences. Its application has thus allowed researchers to identify key factors that have progressively contributed to improving immersion and presence, enabling the creation of more effective and engaging VR systems and environments.

However, the sense of presence is also influenced by the level of cognitive and emotional engagement of the user. In fact, "presence" is not only a technological phenomenon but also a psychological one.

In this regard, the study by Schubert et al. (2001) provided analytical insights into the factors contributing to the sense of presence, proposing a distinction between spatial presence and engagement. In their research, the authors used the PQ, supplemented with items from other questionnaires, to explore the different components of the sense of presence. The results revealed that presence encompasses both a spatial component, related to the perception of being physically in a space, and an engagement component, related to focused attention on the virtual environment. Additionally, a third component emerged, termed "realism," which reflects the user's judgment of the virtual environment's believability compared to the real world (Schubert et al., 2001). Users who feel emotionally and cognitively involved in the virtual experience tend to experience a stronger sense of presence. This suggests that the narrative and content of the VR experience are as crucial as the technology used.

More recently, the study by Cummings and Bailenson (2015) examined the impact of the sense of presence on users' emotional responses, demonstrating that a greater sense of presence is correlated with a more intense emotional response. This is a crucial finding for the design of therapeutic applications in VR, as it suggests that enhancing the sense of presence can increase the effectiveness of virtual reality-based therapies.

1.1.2 *The Transformative Potential of Virtual Reality*

The simulation mechanism that links the mind and VR makes this technology a truly *transformative experience* capable of enhancing and enriching our experience by influencing the sensory, cognitive, and emotional dimensions that characterize it (Gaggioli, 2019). Virtual reality can generate transformative experiences that induce new awareness in individuals, guiding them in revising and restructuring their belief and value systems.

VR environments can evoke the same reactions and emotions as real-world situations, where the sense of presence is strongly correlated with the ability to interact with components of the virtual environment, thus fostering the patient's concentration and engagement. In this context, the generalization of attributions and beliefs can transition from guided VR experiences to everyday real-world situations, effectively leading to a genuine transformation in the patient. The emotional experiences generated by virtual reality are essential in this transformative process because they allow users to develop new awareness of the physical and social context in which they are situated, thereby constructing new meanings.

Virtual environments thus serve as an intermediary between the therapist's office, where protection is maximal, and the external environment, often perceived as threatening. The advantage is that situations, difficulties, events, and consequences can be experienced without any real harm to the patient, who, by virtue of this safety, feels free to explore and experiment.

1.2   Virtual Reality and Psychotherapy: A Foretold Union?

The synergy between virtual reality and the treatment of mental disorders has deep roots, dating back to 1995 when Barbara Rothbaum of the Emory University School of Medicine and Larry Hodges of the Georgia Technology Institute published a pioneering study in the American Journal of Psychiatry. This study investigated the efficacy of gradual VR exposure for the treatment of acrophobia, the fear of heights (Rothbaum et al., 1995). It was an innovative approach aimed at leveraging VR technology to create safe and controlled environments in which patients could gradually confront their fears. The study involved seven weekly sessions, during which participants were gradually exposed to various height situations in a controlled virtual environment. Compared to the control group, participants subjected to gradual VR exposure treatment reported a significant reduction in anxiety, avoidance, and negative attitudes toward heights.

Despite initial skepticism, the results obtained by Rothbaum et al. (1995) encouraged the experimentation of VR in psychotherapeutic treatment to the extent that, since then, the field of VR has made significant strides, expanding not only in the treatment of acrophobia but also in other therapeutic directions. Virtual environments, recreated through new technologies, represent a context of social interaction that allows users to experience emotions and actions, confront their fears and difficulties, and address dysfunctional behaviors. All this occurs within a protected and controlled environment, thus enabling the emergence and reworking of the cognitive material underlying the disorders.

After years of research, virtual reality has established itself as a powerful tool in psychotherapy, thanks to its ability to manipulate imagination and memory—two fundamental aspects that heavily depend on the subject. By wearing a VR headset, the patient is immersed in a simulated situation that can be tailored to address specific psychological disorders. This method not only offers an effective alternative to in vivo exposure but also enhances the patient's perception of safety and control, crucial elements for therapeutic success.

The entire theoretical and methodological framework assumes that individuals immersed in this type of reality experience "presence," the sensation of being in a virtual environment without being aware of the technological mediation, which at a certain point seems to disappear.

Thanks to the flexibility and programmability of the environment, with characteristics varied according to the patient, VR scenes are to be considered a special form of role-playing. The subject, who presumably has not yet found an adequate mode of interaction with the real environment, is offered the opportunity to learn and experiment with new adaptation strategies. This is achieved through exposure to negative stimuli that cause psychological discomfort and subsequent behavioral alterations, with the aim of alleviating the symptoms related to the disorder they suffer from, thereby increasing their level of self-esteem, self-efficacy, and security with contingency management techniques related to panic control, avoidance of compulsive rituals, enactment of self-protective behaviors, and self-deprecating feelings. The user, confronting situations perceived as threatening or anxiety-provoking, will progressively learn to manage them according to an exposure program always agreed upon with the therapist, under their strict supervision.

In this way, the individual is placed in conditions to practically experience that their perception of themselves and the world is not absolute but merely subjective, as it is the result of mental projections, cognitive interpretations, and symbolic representations that can be modified. The virtual environment thus becomes the "secure base," structured and controlled, from which to start exploring, experiencing feelings, imagining, and

reliving those sensations and thoughts (present or past) that psychologically destabilize them.

As we will explore further, the scientific literature has extensively documented the effectiveness of VR in treating various psychological disorders
(Riva et al., 2016). Particularly noteworthy is the attention given to phobia treatment (Botella et al., 2017; Freitas et al., 2021; Demir & Köskün,
2023), which is unsurprising since individuals with specific phobias can be
gradually exposed to the objects of their fears, recreated within a virtual
environment, thus reducing the risk of excessive anxiety that might occur
in real-life situations. This method of gradual exposure is often more
acceptable and less intimidating for patients, thereby enhancing their participation and engagement in the therapeutic process. However, it should
be noted that VR is primarily used as a therapeutic aid, not intending to
replace the fundamental elements of any psychotherapeutic approach, but
rather as a complementary tool that can enhance the efficacy of traditional
treatments. The experimental and clinical protocols that employ VR cannot do without a theoretical orientation and the therapeutic relationship
established with the patient (Freeman et al., 2017). Additionally, it is
important to emphasize that the use of VR in psychotherapy requires specific training for therapists, who must be able to effectively integrate this
tool within their setting and manage any adverse reactions patients may
experience during sessions. Such specific training is crucial, as VR introduces new dynamics into therapy that therapists must be able to manage.
For instance, they must be proficient in configuring and using VR hardware and software, understand its technical limitations, and know how to
adapt the virtual experience to the individual needs of the patients (Botella
et al., 2017). Moreover, therapists must be prepared to recognize and
manage potential adverse reactions experienced by the patient, such as
cybersickness—a form of nausea caused by VR use—or a possible intensification of anxiety that can occur in some patients during virtual exposure
(Rebenitsch & Owen, 2016).

The quality of the therapeutic relationship remains a crucial factor for
the success of the treatment: VR can indeed be a powerful tool, but it will
never replace the therapist, whose empathy and expertise remain pivotal
for a positive treatment outcome. The therapeutic relationship is a central
element in any psychotherapeutic intervention, and numerous studies
have demonstrated that the quality of this relationship is strongly

correlated with positive treatment outcomes (Norcross & Lambert, 2018). Therefore, therapists must balance the use of technology with attention to the emotional and psychological needs of their patients, ensuring that the intervention remains person-centered (Gorini & Riva, 2008).

Box 1.2 The Intervention Protocol in the Treatment of Anxiety Disorders

The intervention protocol adopted for the treatment of anxiety disorders through the use of VR is based on the protocol by Botella et al. (2007), which has been validated through rigorous testing. The methodological core of this intervention is *Systematic Desensitization* (SD), a technique that involves gradual exposure to the phobic stimulus, combined with the prevention of compulsive behavioral responses. SD, recognized as the most established and prevalent practice in behavioral therapy, was originally developed by Joseph Wolpe in the early 1970s (Wolpe, 1973). Predominantly used for treating phobias, SD facilitates the process through which patients are gradually exposed to the feared objects or situations, starting from low and moderate intensity levels, which are increased in subsequent therapeutic sessions. The main objective of this technique is to associate an antagonistic response, such as muscle relaxation or deep breathing, with anxiety-inducing stimuli to attenuate the connection between the anxious reaction and the stimuli that provoke it.

During the treatment, the therapist and the patient jointly develop an inventory of anxiety-inducing stimuli, ranking them according to the degree of discomfort they provoke, with the most disturbing at the top of the list and the least threatening at the bottom. The patient is then guided to initially confront the least anxiety-inducing situations, allowing for a gradual approach to the more problematic stimuli described at the top of the list. This outlines a succession of sessions in which the patient is progressively exposed to phobic situations, with the aim of preventing the compulsive and/or avoidance response, so that a new response can be associated with the phobic stimulus through a process of gradual and controlled conditioning.

1.3   Clinical Applications and Efficacy of VR

Several studies have demonstrated that VR is a valid tool, capable of overcoming some of the common barriers to traditional cognitive-behavioral therapy (CBT), especially in cases where exposure to the triggering factor is necessary. Early studies on the use of VR in psychotherapy provided empirical evidence supporting the effectiveness of this method in treating various anxiety disorders (such as social anxiety disorder, panic disorder, agoraphobia, and specific phobias), disorders related to traumatic and stressful events, and obsessive-compulsive disorder (Pull, 2005; Riva, 2005). Indeed, VR enables experiences that would otherwise be difficult or impossible to achieve in everyday reality. As previously emphasized, VR's ability to create controlled and reproducible environments is particularly advantageous for the gradual and systematic exposure to the factor that triggers the symptoms of the disorder, especially in the case of specific phobias. This is possible because VR technology offers an immersive and interactive experience, allowing patients to confront anxiety-provoking situations in a controlled and safe environment.

VR-mediated therapy progressively desensitizes the individual from their anxieties, enabling them to manage emotions through a different approach. This technique, reminiscent of cognitive-behavioral *flooding*, is implemented with advanced technological tools that significantly increase the control exercised by the therapist. The components of the virtual environment are entirely under the control of the therapist, who can modulate the level of difficulty presented to the patient, taking on the role of mediator between the real world and the virtual one.

This approach is particularly advantageous for the treatment of anxiety disorders, such as specific phobias and panic disorder. For example, VR support can simulate situations like boarding an airplane, speaking in front of an audience, or gradually and safely exposing the subject to feared stimuli (Maples-Keller et al., 2017). This methodology, known as *Virtual Reality Exposure Therapy* (VRET), was extensively discussed in a study by Krijn et al. (2004), which highlighted how it integrates computer graphics, body tracking devices, and other sensory technologies to immerse patients in virtual environments, thus facilitating more effective systematic desensitization.

Specific reviews and meta-analyses, some of which have been published in the Cochrane Library, have examined studies on specific phobias such as fear of flying, agoraphobia, fear of driving, claustrophobia, and arachnophobia (Parsons & Rizzo, 2008; Botella et al., 2017). In most cases, VRET has shown more than promising results, proving to be as effective as traditional in vivo exposure techniques, but with the additional advantage of being customizable and modifiable according to the individual needs of the patient, allowing for a tailored therapeutic approach that can lead to better clinical outcomes (Freeman et al., 2018). Recent studies have indicated that, in some cases, VRET has even surpassed the effectiveness of traditional exposure therapy, reducing anxiety symptoms and increasing patients' sense of security and control in significantly shorter times (Carl et al., 2019). This is likely due to the fact that VR can be used effectively to create immersive experiences that are otherwise difficult to achieve (Parsons & Rizzo, 2008). For instance, a recent study by Tsamitros et al. (2023) demonstrated that VRET is as effective as in vivo exposure for the treatment of specific phobias and agoraphobia with panic disorder, and significantly more effective compared to control groups and placebo for the treatment of social phobia (Tsamitros et al., 2023).

Significant results have also been achieved in the treatment of disorders related to traumatic and stressful events, such as post-traumatic stress disorder (PTSD). Notably, one of the pioneers in this field is Dr. Albert Rizzo from the Institute for Creative Technology at the University of Southern California, who has used VRET for the treatment of PTSD in veteran soldiers who fought in Iraq, recreating stressful situations using a VR video game that simulated various military-related scenes (Rizzo et al., 2005). The simulation was based on the idea of re-experiencing all the emotions that had caused the trauma, but this time safely, in the therapist's office. The difference with other techniques previously used is the ability to immerse oneself in the past experience and do so gradually, excluding anything the patient is not yet ready to face, providing a safe and controlled environment for the processing of trauma.

The use of VR for the treatment of PTSD has received increasing attention and support in the most recent scientific literature. Several studies have explored the efficacy of VR in this context, highlighting the benefits of VRET as a therapeutic tool. Following the research line cited by Rizzo et al. (2005), a recent review examined the use of virtual reality techniques for the rehabilitation of military veterans with PTSD (Volovik et al., 2023). This review included 75 articles published between 2017 and 2022, analyzing the therapeutic mechanisms of VRET and protocols combined with other therapies, such as pharmacotherapy and transcranial magnetic stimulation. The results showed that the inclusion of VRET in PTSD rehabilitation programs significantly improves outcomes due to the enhanced sense of presence and greater customization of the experience. Therefore, VRET has proven to be a valid, controlled, and cost-effective alternative for the treatment of PTSD in combatants, including those more resistant to conventional therapies.

1.3.1   VR Applications for Other Disorders

The application of VR is also scientifically supported for the treatment of other disorders, such as obsessive-compulsive disorder (OCD), body dysmorphic disorder (BDD), and schizophrenia.

In the case of using virtual reality for the treatment of OCD, the work of Clark et al. (1998) laid the groundwork, proposing vicarious exposure to anxiety-inducing stimuli (often related to contamination) combined with the prevention of compulsive rituals (Clark et al., 1998). A recent study suggested using selected virtual environments as reference stimuli to obtain an assessment and formulate an interactive measure of compulsive control behaviors (Kim et al., 2010).

Although the use of virtual reality with patients suffering from schizophrenia is relatively recent, some studies demonstrate that it allows interesting applications for both assessment and treatment in a controlled environment. This tool enables the reproduction of environmental and social situations that stimulate the subject in a manner similar to the real context; furthermore, the intensity and duration of the virtual experience can be modulated according to the subject's needs (La Barbera et al., 2010). It can also reproduce emotional and social situations typical of interpersonal relationships (Kim et al., 2010). Virtual environments, as in the treatment of phobias, allow patients to be exposed to their persecutory fears and test their beliefs about what is perceived as threatening.

Another use of VR is to teach coping strategies to adopt in various social situations when psychotic symptoms occur. For example, virtual reality has been applied in role-playing games to stimulate these patients' interpersonal skills, improving their conversational abilities and self-confidence (Park et al., 2011).

The recent systematic review by Lan et al. (2023) examined the use of VR and augmented reality (AR) for psychotic disorders, demonstrating that adding VR therapies to traditional treatments significantly improves clinical outcomes, including medication adherence and rehabilitation, and is both acceptable and safe for patients (Lan et al., 2023).

In another study, Miranda et al. (2024) developed an immersive serious VR game for treating the negative symptoms of schizophrenia, such as apathy, social withdrawal, and lack of motivation, which can be difficult to address with medication alone (Miranda et al., 2024). The study explored innovative solutions to complement conventional treatments for schizophrenia, particularly for symptoms that are less responsive to medication. The study involved 21 healthy subjects who participated in tests on a VR game featuring a tutorial and two levels of difficulty. The objective of the game was to improve specific cognitive and social functions through immersive interaction. Preliminary results were promising, indicating that VR could be a valuable complement to conventional treatments for schizophrenia. However, the authors emphasized the need for further research to develop more effective and accessible VR solutions, reducing costs and expanding treatment reach (Miranda et al., 2024).

Nonetheless, the main limitation of virtual applications with patients suffering from severe conditions is the stability of reality testing, which characterizes the acute phase of the illness. Each schizophrenic patient can have different symptoms and reactions to VR therapy. Adapting VR programs to individual needs can be complex and require additional resources. The study by Rault et al. (2022) examined the therapeutic use of VR for relaxation in patients with schizophrenia, showing that patients tolerated VR well, resulting in decreased anxiety. However, personalizing the experience for each patient was crucial to ensure treatment efficacy (Rault et al., 2022). VR content must be carefully designed and customized for individual users, especially those with severe disorders like schizophrenia: this requires a deep understanding of the patient's needs and capabilities, as well as resources to continuously develop and adapt these environments.

Moreover, in the case of severe disorders, the main risk is that, in some cases, symptoms such as hallucinations and delusions may be exacerbated using VR. In a randomized controlled study on the effect of VR on cognitive impairment and clinical symptoms among schizophrenic patients, Sharma and Lippke (2022) emphasized the need for closer and continuous monitoring of patients to avoid negative effects, preventing possible exacerbations of symptoms (Sharma & Lippke, 2022).

1.3.2   Use of VR in the Treatment of Eating Disorders

Several studies have explored the effectiveness of VR in the treatment of eating disorders, highlighting how patients experience a high sense of presence when exposed to virtual environments, as confirmed by various research (Riva et al., 2021; Ferrer-García & Gutiérrez-Maldonado, 2012; Perpiñá et al., 2003). Since the 1990s, virtual reality has been integrated into therapeutic protocols for eating disorders, complementing pharmacological treatments and cognitive therapies (Riva et al., 2004). In particular, VR has been used as a tool to correct distorted body image perceptions, allowing for a comparison between the patient's subjective body perception and their actual physical shape, measured objectively (Riva, 1998).

Recent studies, including that of Porras-Garcia et al. (2021), have demonstrated that combining VR with psychotherapy, through the use of three-dimensional avatars, is particularly effective in mitigating the fear of gaining weight (FGW) and body image disturbances (BIDs) associated with anorexia (Porras-Garcia et al., 2021). The benefits of this approach were also observed in a three-month follow-up, with a significant reduction in eating disorder symptoms and an increase in body mass index. These results were further confirmed by a more recent study by Behrens et al. (2023), which emphasizes that exposure to representations of healthy and normal-weight bodies can significantly improve body perception and eating behaviors in patients with anorexia (Behrens et al., 2023).

Other studies have examined how virtual food environments, known as "food-cues," influence the activation, maintenance, and/or relapse mechanisms associated with bulimia nervosa and binge eating disorder. Exposure to virtual food can elicit reactions similar to those triggered by exposure to real food, as these virtual environments are capable of evoking emotional, cognitive, and behavioral responses in subjects using virtual reality (Ferrer-García et al., 2014; Pla-Sanjuanelo et al., 2015). Moreover, Ferrer-García et al. (2014) and Ferrer-García and Gutiérrez-Maldonado (2012) confirmed that food environments featuring high-calorie foods and certain social situations can generate high levels of subjective discomfort. According to Perpiñá et al. (2013), emotional reactivity to virtual reality tends to increase proportionally with the perceived sense of presence in the virtual environment (Perpiñá et al., 2013).

Although not recognized as diagnostic categories in the DSM-5-TR, VR interventions concerning obesity have demonstrated their effectiveness. Specific studies, such as that of Manzoni et al. (2015), found that virtual reality can help obese patients develop a more realistic perception of their bodies and assist those with eating disorders in reducing their body dissatisfaction (Manzoni et al., 2015).

1.3.3 VR and Psychotherapy: What Are the Limitations?

VR has found an increasingly wide range of applications in psychotherapy, but the available literature and practical implementation opportunities reveal some significant limitations. One of the primary obstacles to adopting this technology is the high level of technological competence required to use VR devices (Riva & Gaggioli, 2019). These skills are generally acquired outside of traditional training paths, making the technology less accessible in the clinical context. Additionally, the high installation costs and the need for ongoing technical assistance can make the use of VR prohibitive in care centers and especially in private practices (Riva, 2012).

Another issue concerns the specific technological requirements, which, if not met, often represent a major obstacle to the effectiveness of VR. The quality of the virtual experience heavily depends on the sophistication of the hardware and software used: less advanced VR devices can produce low-quality images, delays, or connectivity problems, which could compromise the immersion and effectiveness of the treatment. Moreover, the necessity of specific and often expensive devices can limit the accessibility of such treatments for patients with limited resources (Riva & Gaggioli, 2019).

Lastly, attention must be paid to the phenomenon known as *cybersickness*, which affects a minority of subjects and includes symptoms such as dizziness, nausea, fatigue, and disorientation. These effects are the result of a discrepancy between the visually perceived movement and the absence of corresponding physical movement, which is not detected by the vestibular system (Kourtesis et al., 2024). Although research in this area is still ongoing, some studies have shown significant improvements with repeated exposure to VR, suggesting a progressive decrease in the intensity of symptoms (Kourtesis et al., 2024).

1.3.4 Reflections on Potential Future Developments

In the realm of modern therapeutic practices, VR emerges as an intriguing expansion of the intervention repertoire, particularly where traditional psychotherapy struggles to achieve optimal results, thereby increasing the potential pool of users who can benefit from psychotherapeutic intervention.

This innovative technology stands out for its ability to offer a complete sensory experience, which can be difficult to replicate in a conventional therapeutic scenario, especially for those individuals who, due to their state of inhibition, find it challenging to generate vivid and engaging mental images.

Specifically, it has been shown that the introduction of VR proves particularly effective not only for individuals who are reluctant to participate in in vivo therapies but also for patients with reduced imaginative proclivity, a condition that could hinder traditional psychotherapeutic techniques that rely on guided visualization or similar methods (North et al., 1998; Vincelli & Molinari, 1998). A recent systematic review, besides demonstrating the effectiveness of VRET in reducing symptoms of specific phobias and maintaining these results in the long term, highlighted how this method is particularly useful for patients who have difficulty imagining anxiety-provoking scenes or are too phobic to face real situations (Demir & Köskün, 2023). VR, through the use of immersive and controlled environments, allows for the configuration of highly specific and detailed personalized scenarios, capable of replicating the anxiety-provoking or stressful situations that the patient will subsequently face. In this way, the subject is placed in a position to experience and process their reactions in a safe and reversible context, thereby enhancing their ability to handle similar contexts in real life. This approach is a form of controlled exposure, where the patient, assisted by a therapist, can gradually get accustomed to stressogenic stimuli, progressively reducing the anxious response.

The potential of VR is not limited to improving existing treatments but also opens new avenues for psychotherapy. Another innovative aspect of great interest to the scientific community is the capability of VR to facilitate self-managed treatments. As highlighted by Shin et al. (2021), the use of mobile applications for self-managing treatments has shown significant improvements in panic disorder symptoms compared to control groups. This mode of intervention offers patients greater autonomy and flexibility, potentially transforming the traditional therapeutic paradigm. However, despite the undeniable benefits, there are several risks and challenges associated with the use of VR in self-management contexts that merit consideration.

More precisely, the absence of the therapist involves two significant risk factors that seem almost inevitable in the case of self-managed treatments: lack of personalization and isolation from technical support.

In the first case, it is important to consider that VR applications often use pre-packaged scenarios that may not perfectly fit the individual needs of the patient or the specificity of their symptoms. During VR treatment, personalization becomes crucial so that the intervention is adapted to the unique circumstances of each individual.

In the second case, the lack of regular interaction with a therapist can prevent adequate monitoring of the patient's progress and response to treatment. This can be particularly problematic in crisis situations or when new symptoms emerge. Technical support and clinical supervision are essential to promptly address any issues or adjust the treatment based on emerging needs.

Isolation from technical support can also lead to improper use of VR technology, which could not only reduce the effectiveness of the treatment but also cause undesirable side effects. For instance, prolonged or inadequately monitored exposure to stressful scenarios can increase anxiety instead of reducing it. Additionally, the lack of professional feedback can prevent the patient from recognizing and adequately managing emerging symptoms, increasing the risk of worsening the condition.

Besides the aforementioned risks, the long-term effects of using VR as a self-managed treatment tool remain unclear. Long-term research is still limited, and it is not yet fully understood whether the improvements observed during VR use persist over time once the intervention is completed (Freeman et al., 2017).

The rapid pace at which virtual reality systems are developing, increasing their accessibility each time, thus calls for conscious and ethical reflection on the potential misuse of such technology, especially considering that subjects might exhibit very intense reactions to simulated environments.

1.4   Exploring the Impact of VR on Emotions and Cognitive Functions

The use of VR has opened new frontiers in multiple sectors, extending beyond its initial application in video game entertainment to fields such as education, medicine, and psychology. Specifically, VR has been employed in the treatment of various mental disorders with promising results. Numerous studies have demonstrated the effectiveness of VR in enhancing cognitive functions and managing emotions through the use of controlled protocols, highlighting significant effects on the patients' minds and behaviors (Freeman et al., 2017; Valmaggia et al., 2016).

The revolutionary impact of VR in the field of mental health is undeniable, transforming some of the most well-known therapeutic approaches. One of the primary applications of VR is cognitive rehabilitation. By creating highly realistic and immersive environments, patients can practice complex cognitive tasks in a safe and controlled context, consequently improving their performance (Riva et al., 2016).

An example is highlighted in the study by Argüero-Fonseca et al. (2023), which explored the effect of a VR-based mindfulness intervention on a sample of adolescents. The intervention utilized the TRIPP application on Oculus Rift S headsets and showed significant improvements in the participants' divided attention and working memory capabilities after just 21 days. This study underscores how VR, by implementing immersive and interactive environments, stimulates cognitive functions more directly and intensively than traditional approaches, especially in areas such as attention and working memory, which are crucial for academic and professional success (Argüero-Fonseca et al., 2023).

Another domain where VR has emerged as a transformative technology is education, offering new pathways for immersive and interactive learning experiences. The primary advantage of using VR lies in its ability to introduce a mode of learning that transcends conventional reality, allowing users to immerse themselves in three-dimensional landscapes that would otherwise be inaccessible, dynamically interacting with the virtual environment. These features of presence, interactivity, and immersion are fundamental for understanding its potential impact on student engagement across various educational contexts (Chen et al., 2024; Merchant et al., 2014).

Simulations of real situations for training purposes represent a key application of VR. VR simulations can recreate complex scenarios where students must apply their knowledge and skills to solve problems. For example, in the medical field, future surgeons can perform virtual operations, improving their dexterity and decision-making abilities under stress. Similarly, pilots can use VR simulations to train for handling in-flight emergencies (Lin et al., 2024).

VR can also stimulate memory by creating environments rich in visual and spatial stimuli. For instance, exploring a virtual city and having to remember the route back to a starting point helps improve spatial memory. This type of training is particularly useful for individuals with memory deficits (Cimadevilla et al., 2023). Moreover, the immersive and often playful nature of VR increases student engagement and motivation: integrating game elements into learning programs can make educational activities more enjoyable and stimulating. This increased engagement can lead to greater attention and concentration, enhancing the effectiveness of learning (Huang et al., 2010; Makransky & Lilleholt, 2018).

The research conducted by Lin et al. (2024) critically explores how VR influences students' cognitive, behavioral, and affective engagement, demonstrating that VR use can significantly enhance student engagement, proving particularly effective for those with Specific Learning Disabilities. However, introducing VR into middle school education presents challenges, including the difficulty for the educational system to keep pace with VR technological advancements, the increased demand for students' digital skills, and the insufficient capacity of teachers in using VR. To overcome these obstacles, the authors suggest that educational policymakers provide training and technical support to teachers to ensure full mastery and integration of VR, thereby increasing student engagement and teaching effectiveness. These recommendations are supported by studies showing that well-designed VR environments can greatly enrich the educational experience, improving students' attention, comprehension, and information retention (Lin et al., 2024).

1.5   Some Brief Final Considerations

At its deepest essence, VR serves as a metaphor for the human potential to extend and transcend one's own boundaries, offering not only a refuge from occasionally oppressive realities but also a space to explore and engage, benefiting from the possibility of going beyond space and time.

The ability of VR to simulate realistic and controlled environments, as demonstrated in the treatment of psychological disorders such as anxiety, PTSD, and phobias, represents a profound shift from the traditional paradigm of psychotherapy. It is no longer the individual adapting to the world, but the world shaping itself around individual needs, offering a safe place to confront and process trauma at a personal and controlled pace. This reversibility of control, from the world to the individual, constitutes a form of psychological empowerment that promotes a sense of agency and self-efficacy.

Beyond therapy, VR emerges as a fertile ground for education and training. Immersive and interactive environments enhance learning and knowledge retention, creating situations where students are not mere spectators but active protagonists in their own educational process. This shift from a passive to an active learning scenario underscores a change in the perception of learning: from obligation to an opportunity for personal exploration and development.

These applications of VR are not merely technological innovations but reflect a broader cultural and philosophical transformation in the relationship between the individual and reality. The fact that we can create alternative worlds to confront our fears or develop new skills raises essential questions about the meaning of reality and how we define human experience. In this sense, VR acts as a mirror of our deepest aspirations, reflecting our incessant quest for understanding, control, and ultimately transcending our bodily and psychological limits.

Finally, the progressive fusion of VR with other technologies such as artificial intelligence and machine learning promises to make these environments increasingly responsive and personalized, adapting in real time to the emotional and cognitive needs of the user. This development raises profound questions about the future of identity and personal autonomy in a world where the boundaries between the real and the virtual become increasingly blurred.

REFERENCES

Argüero-Fonseca, A., Martínez-Soto, J., Aguirre-Ojeda, D. P., Pérez-Pimienta, D., & Marchioro, D. M. (2023). Effects of a protocol of environmental psychological restauration with virtual reality on indicators of demotivation in children. *Journal of Population Therapeutics & Clinical Pharmacology, 30*(17), 821–828. https://doi.org/10.53555/jptcp.v30i18.3182

Behrens, S. C., Tesch, J., Sun, P. J., Starke, S., Black, M., Schneider, H., Pruccoli, J., Zipfel, S., & Giel, K. E. (2023). Virtual reality exposure to a healthy weight body is a promising adjunct treatment for anorexia nervosa. *Psychotherapy and Psychosomatics, 92*(3), 170–179. https://doi.org/10.1159/000530932

Botella, C., Baños, R. M., García-Palacios, A., & Quero, S. (2017a). Virtual reality and other realities. In S. G. Hofmann & G. J. Asmundson (Eds.), *The science of cognitive behavioral therapy* (pp. 551–590). Elsevier Academic Press.

Botella, C., Fernández-Álvarez, J., Guillén, V., & García-Palacios, A. (2017b). Recent Progress in virtual reality exposure therapy for phobias: A systematic review. *Current Psychiatry Reports, 19*(42), 1–13. https://doi.org/10.1007/s11920-017-0788-4

Botella, C., García-Palacios, A., Villa, H., Baños, R. M., Quero, S., Alcañiz, M., & Riva, G. (2007). Virtual reality exposure in the treatment of panic disorder and agoraphobia: A controlled study. *Clinical Psychology and Psychotherapy, 14*(3), 164–165. https://doi.org/10.1002/cpp.524

Botella, C., Serrano, B., Baños, R. M., & Garcia-Palacios, A. (2015). Virtual reality exposure-based therapy for the treatment of post-traumatic stress disorder: A review of its efficacy, the adequacy of the treatment protocol, and its acceptability. *Neuropsychiatric Disease and Treatment, 11*, 2533–2545. https://doi.org/10.2147/NDT.S89542

Carl, E., Stein, A. T., Pogue, J. R., Rothbaum, B. O., Emmelkamp, P., Asmundson, G. J., Carlbring, P., & Powers, M. B. (2019). Virtual reality exposure therapy for anxiety and related disorders: A meta-analysis of randomized controlled trials. *Journal of Anxiety Disorders, 61*, 27–36. https://doi.org/10.1016/j.janxdis.2018.08.003

Chen, J., Fu, Z., Liu, H., & Wang, J. (2024). Effectiveness of virtual reality on learning engagement: A meta-analysis. *International Journal of Web-Based Learning and Teaching Technologies, 9*(1), 1–14. https://doi.org/10.4018/IJWLTT.334849

Cimadevilla, J. M., Nori, R., & Piccardi, L. (2023). Application of Virtual Reality in Spatial Memory. *Brian Sciences, 13*(1621), 1–4. https://www.mdpi.com/journal/brainsci

Clark, A., Kirkby, K. C., Daniels, B. A., & Marks, I. M. (1998). A pilot study of computer-aided vicarious exposure for obsessive-compulsive disorder. *Australian & New Zealand Journal of Psychiatry, 32*(2), 268–275. https://doi.org/10.3109/00048679809062738

Cummings, J. J., & Bailenson, J. N. (2015). How immersive is enough? A meta-analysis of the effect of immersive technology on user presence. *Media Psychology, 19*(2), 272–309. https://doi.org/10.1080/15213269.2015.1015740

Demir, M., & Köskün, T. (2023). Efficacy of virtual reality exposure therapy in the treatment of specific phobias: A systematic review. *Psikiyatride Güncel Yaklaşımlar, 15*(4), 562–576. https://doi.org/10.18863/pgy.1192543

Ellis, S. R. (1994). What are virtual environments? *IEEE Computer Graphics and Applications, 14*(1), 17–22. https://doi.org/10.1109/38.250914

Ferrer-García, M., & Gutiérrez-Maldonado, J. (2012). The use of virtual reality in the study, assessment, and treatment of body image in eating disorders and non-clinical samples: a review of the literature., *9*(1), 1–11. https://doi.org/10.1016/j.bodyim.2011.10.001

Ferrer-García, M., Gutiérrez-Maldonado, J., Pla-Sanjuanelo, J., Riva, G., Andreu-Gracia, A., Dakanalis, A., Fernandez-Aranda, F., Forcano, L., Ribas-Sabaté, J., Riesco, N., Rus-Calafel, M., & Sánchez, I. (2014). Development of a VR application for binge eating treatment: Identification of contexts and cues related to bingeing behavior in Spanish Italian patients. *Studies in Health Technology and Informatics, 71-75.* https://doi.org/10.3233/978-1-61499-401-5-71

Freeman, D., Haselton, P., Freeman, J., Spanlang, B., Kishore, S., Albery, E., Albery, E., Denne, M., Brown, P., Slater, M., & Nickless, A. (2018). Automated psychological therapy using immersive virtual reality for treatment of fear of heights: A single-blind, parallel-group, randomised controlled trial. *Lancet Psychiatry, 5*(8), 625–632. https://doi.org/10.1016/S2215-0366(18)30226-8

Freeman, D., Reeve, S., Robinson, A., Ehlers, A., Clark, D. M., Spanlang, B., & Slater, M. (2017). Virtual reality in the assessment, understanding, and treatment of mental health disorders. *Psychological Medicine, 47*(14), 1–8. https://doi.org/10.1017/S003329171700040X

Freitas, R., Velosa, V. H., Abreu, L. T., Jardim, R. L., Santos, J. A., Peres, B., & Campos, P. (2021). Virtual reality exposure treatment in phobias: A systematic review. *Psychiatric Quarterly, 92*(4), 1–26. https://doi.org/10.1007/s11126-021-09935-6

Gaggioli, A. (2019). What is it like to be a tree? The transformative potential of virtual reality. *Cyberpsychology, Behavior, and Social Networking, 22*(3), 232–232. https://doi.org/10.1089/cyber.2019.29144.csi

Gorini, A., & Riva, G. (2008). Virtual reality in anxiety disorders: The past and the future. *Expert Review of Neurotherapeutics, 8*(2), 215–233. https://doi.org/10.1586/14737175.8.2.215

Huang, H.-M., Rauch, U., & Liaw, S. (2010). Investigating learners' attitudes toward virtual reality learning environments: Based on a constructivist approach. *Computers & Education, 55*(3), 1171–1182. https://doi.org/10.1016/j.compedu.2010.05.014

Kim, K., Kim, S. I., Cha, K. R., Park, J., Rosenthal, M. Z., Kim, J. J., Han, K., Kim, I. Y., & Kim, C. H. (2010). Development of a computer-based behavioral assessment of checking behavior in obsessive-compulsive disorder. *Comprehensive Psychiatry, 86*–93. https://doi.org/10.1016/j.comppsych.2008.12.001

Kourtesis, P., Papadopoulou, A., & Roussos, P. (2024). Cybersickness in virtual reality: The role of individual differences, its effects on cognitive functions and motor skills, and intensity differences during and after immersion. *Virtual Worlds, 3*(1), 62–93. https://doi.org/10.3390/virtualworlds3010004

Krijn, M., Emmelkamp, P., Ólafsson, R. P., & Biemond, R. (2004). Virtual reality exposure therapy of anxiety disorders: A review. *Clinical Psychology Review, 24*(3), 259–281. https://doi.org/10.1016/J.CPR.2004.04.001

Krueger, M. W. (1991). *Artificial reality* (2nd ed.). Addison-Wesley Publishing.

La Barbera, D., Sideli, L., & La Paglia, F. (2010). Schizophrenia and virtual reality: A review of clinical applications. *Italian Journal of Psychopathology*, *16*, 78–86.

Lan, L., Sikov, J., Lejeune, J., Ji, C., Brown, H., Bullock, K., & Spencer, A. E. (2023). A systematic review of using virtual and augmented reality for the diagnosis and treatment of psychotic disorders. *Current Treatment Options in Psychiatry*, *10*, 87–107. https://doi.org/10.1007/s40501-023-00287-5

Lin, X. P., Li, B. B., Yao, Z. N., Yang, Z., & Zhang, M. (2024). The impact of virtual reality on student engagement in the classroom–a critical review of the literature. *Frontiers in Psychology*, *15*, 1–8. https://doi.org/10.3389/fpsyg.2024.1360574

Lindner, P., Miloff, A., Hamilton, W., Reuterskiöld, L., Andersson, G., Powers, M. B., & Carlbring, P. (2017). Creating state of the art, next-generation virtual reality exposure therapies for anxiety disorders using consumer hardware platforms: Design considerations and future directions. *Cognitive Behaviour Therapy*, *46*(5), 404–420. https://doi.org/10.1080/16506073.2017.1280843

Makransky, G., & Lilleholt, L. (2018). A structural equation modeling investigation of the emotional value of immersive virtual reality in education. *Educational Technology Research and Development*, *66*(3), 1141–1164. https://doi.org/10.1007/s11423-018-9581-2

Manzoni, G. M., Cesa, G. L., Bacchetta, M., Castelnuovo, G., Conti, S., Gaggioli, A., Mantovani, F., Molinari, E., Cárdenas-López, G., & Riva, G. (2015). Virtual reality–enhanced cognitive–behavioral therapy for morbid obesity: A randomized controlled study with 1 year follow-up. *Cyberpsychology, Behavior and Social Networking*, 1–7. https://doi.org/10.1089/cyber.2015.0208

Maples-Keller, J. L., Bunnell, B. E., Kim, S.-J., & Rothbaum, B. O. (2017). The use of virtual reality Technology in the Treatment of anxiety and other psychiatric disorders. *Harvard Review of Psychiatry*, *25*(3), 103–113. https://doi.org/10.1097/HRP.0000000000000138

Merchant, Z., Goetz, E. T., Cifuentes, L., Keeney-Kennicutt, W., & Davis, T. J. (2014). Effectiveness of virtual reality-based instruction on students' learning outcomes in K-12 and higher education: A meta-analysis. *Computers & Education*, *70*, 29–40. https://doi.org/10.1016/j.compedu.2013.07.033

Miranda, B., Rego, P. A., Romero, L., & Moreira, P. M. (2024). Application of immersive VR serious games in the treatment of schizophrenia negative symptoms. *Computers*, *13*(2), 1–17. https://doi.org/10.3390/computers13020042

Norcross, J. C., & Lambert, M. J. (2018). Psychotherapy relationships that work III. *Psychotherapy*, *55*(4), 303–315. https://doi.org/10.1037/pst0000193

North, M. M., North, S. M., & Coble, J. K. (1998). Virtual reality therapy: An effective treatment for phobias. *Studies in Health Technology and Informatics, 58*, 112–119. https://doi.org/10.3233/978-1-60750-902-8-112

Park, K.-M., Ku, J., Choi, S.-H., Jang, H.-J., Park, J.-Y., Kim, S. I., & Kim, J.-J. (2011). A virtual reality application in role-plays of social skills training for schizophrenia: A randomized, controlled trial author links open overlay panel. *Psychiatry Research, 189*(2), 166–172. https://doi.org/10.1016/j.psychres.2011.04.003

Parsons, T. D., & Rizzo, A. (2008). Affective outcomes of virtual reality exposure therapy for anxiety and specific phobias: A meta-analysis. *Journal of Behavior Therapy and Experimental Psychiatry, 39*(3), 250–261. https://doi.org/10.1016/j.jbtep.2007.07.007

Perpiñá, C., Botella, C., & Baños, R. M. (2003). Virtual reality in eating disorders. *European Eating Disorders Review, 261*–278. https://doi.org/10.1002/erv.520

Perpiñá, C., Roncero, M., Fernández-Aranda, F., Jiménez-Múrcia, S., Forcano, L., & Sánchez, I. (2013). Clinical validation of a virtual environment for normalizing eating patterns in eating disorders. *Comprehensive Psychiatry, 54*(6), 680–686. https://doi.org/10.1016/j.comppsych.2013.01.007

Pla-Sanjuanelo, J., Ferrer-García, M., Gutiérrez-Maldonado, J., Riva, G., Andreu-Gracia, A., Dakanalis, A., Fernandez-Aranda, F., Forcano, L., Ribas-Sabaté, J., Riesco, N., Rus-Calafell, M., & Sánchez, I. (2015). Identifying specific cues and contexts related to bingeing behavior for the development of effective virtual environments. *Appetite, 87*(1), 81–89. https://doi.org/10.1016/j.appet.2014.12.098

Porras-Garcia, B., Ferrer-Garcia, M., Serrano-Troncoso, E., Carulla-Roig, M., Soto-Usera, P., Miquel-Nabau, H., Fernandez-Del Castillo Olivares, L., Marnet-Fiol, R., de la Montaña, S.-C. I., Borszewski, B., Díaz-Marsá, M., & Borszewski, B. (2021). AN-VR-BE. A randomized controlled trial for reducing fear of gaining weight and other eating disorder symptoms in anorexia nervosa through virtual reality-based body exposure. *Journal of Clinical Medicine, 10*(682), 1–23. https://doi.org/10.3390/jcm10040682

Pull, C. B. (2005). Current status of virtual reality exposure therapy in anxiety disorders. *Current Opinion in Psychiatry, 18*(1), 7–14.

Rault, O., Lamothe, H., & Pelissolo, A. (2022). Therapeutic use of virtual reality relaxation in schizophrenia: A pilot study. *Psychiatry Research, 309.* https://doi.org/10.1016/j.psychres.2022.114389

Rebenitsch, L., & Owen, C. (2016). Review on cybersickness in applications and visual displays. *Virtual Reality, 20*, 101–125. https://doi.org/10.1007/s10055-016-0285-9

Riva, G. (1998). Virtual environments in neuroscience. *IEEE Transactions on Information Technology in Biomedicine, 2*(4), 275–281. https://doi.org/10.1109/4233.737583

Riva, G. (2005). Virtual reality in psychotherapy: Review. *Cyberpsychology & Behavior*, 220–230. https://doi.org/10.1089/cpb.2005.8.220

Riva, G. (2006). Virtual reality. In M. Akay (Ed.), *Encyclopedia of biomedical engineering*. John Wiley & Sons.

Riva, G. (2012). *Psicologia dei nuovi media. Azione, presenza, identità e relazione nei media digitali e nei social media*. Il Mulino.

Riva, G. (2018). The neuroscience of body memory: From the self through the space to the others. *Cortex, 104*, 241–260. https://doi.org/10.1016/j.cortex.2017.07.013

Riva, G., Bacchetta, M., Cesa, G., Conti, S., & Molinari, E. (2004). The use of VR in the treatment of eating disorders. In I. G. Riva, C. Botella, P. Legeron, & G. Optale (Eds.), *Cybertherapy: Internet and virtual reality as assessment and rehabilitation tools for clinical psychology and neuroscience* (Vol. 99, pp. 121–163). IOS Press.

Riva, G., Baños, R. M., Botella, C., Mantovani, F., & Gaggioli, A. (2016). Transforming experience: The potential of augmented reality and virtual reality for enhancing personal and clinical change. *Frontiers in Psychiatry, 7*, 1–14. https://doi.org/10.3389/fpsyt.2016.00164

Riva, G., Davide, F., & Ijsselsteijn, W. A. (2003). *Being there: Concepts, effects and measurements of user presence in synthetic environments*. IOS Press.

Riva, G., & Gaggioli, A. (2019). *Realtà virtuali: Gli aspetti psicologici delle tecnologie simulative e il loro impatto sull'esperienza umana*. Giunti Psychometrics.

Riva, G., Malighetti, C., & Serino, S. (2021). Virtual reality in the treatment of eating disorders. *Clinical Psychology & Psychotherapy, 28*(3), 477–488. https://doi.org/10.1002/cpp.2622

Rizzo, A., Pair, J., McNerney, P. J., Eastlund, E., Manson, B., Gratch, J., Hill, R., & Swartout, B. (2005). Development of a VR therapy application for Iraq war military personnel with PTSD. *Studies in Health Technology and Informatics, 111*, 407–413. https://doi.org/10.1007/978-3-319-66192-6_27

Rothbaum, B. O., Hodges, L. F., Kooper, R., Opdyke, D., Williford, J. S., & North, M. M. (1995). Effectiveness of computer-generated (virtual reality) graded exposure in the treatment of acrophobia. *American Journal of Psychiatry, 152*(4), 626–628. https://doi.org/10.1176/ajp.152.4.626

Rubio-Tamayo, J. L., Barrio, M. G., & García, F. G. (2017). Immersive environments and virtual reality: Systematic review and advances in communication, interaction and simulation. *Multimodal Technologies and Interaction, 1*(4), 21. https://doi.org/10.3390/mti1040021

Sastry, L., & Boyd, D. (1998). Virtual environments for engineering applications. *Virtual Reality. Research, development and applications, 3*(4), 235–244. https://doi.org/10.1007/BF01408704

Schubert, T., Friedmann, F., & Regenbrecht, H. (2001). The experience of presence: Factor analytic insights. *Presence: Teleoperators and Virtual Environments, 10*(3), 266–281. https://doi.org/10.1162/105474601300343603

Schultheis, M. T., & Rizzo, A. (2001). The application of virtual reality technology in rehabilitation. *Rehabilitation Psychology, 46*(3), 296–311. https://doi.org/10.1037/0090-5550.46.3.296

Sharma, V., & Lippke, R. L. (2022). Effect of virtual reality on cognitive impairment and clinical symptoms among patients with schizophrenia in the remission stage: A randomized controlled trial. *Brain Sciences, 12*(11), 1572–1572. https://doi.org/10.3390/brainsci12111572

Shin, B., Oh, J., Kim, B., Kim, H., Kim, H., Kim, S., & Kim, J. (2021). Effectiveness of Self-Guided Virtual Reality–Based Cognitive Behavioral Therapy for Panic Disorder: Randomized Controlled Trial. *JMIR Mental Health, 8*(11). https://doi.org/10.2196/30590

Slater, M. (2009). Place illusion and plausibility can lead to realistic behaviour in immersive virtual environments. *Philosophical Transactions of the Royal Society of London. Series B, Biological Sciences, 364*(1535), 3549–3557. https://doi.org/10.1098/rstb.2009.0138

Slater, M., & Sanchez-Vives, M. V. (2016). Enhancing our lives with immersive virtual reality. *Frontiers in Robotics and AI*, Art. 74. https://doi.org/10.3389/frobt.2016.00074

Steuer, J. (1992). Defining virtual reality: Dimensions determining telepresence. *Journal of Communication, 42*(4), 73–93. https://doi.org/10.1111/j.1460-2466.1992.tb00812.x

Tsamitros, N., Beck, A., Sebold, M., Schouler-Ocak, M., Bermpohl, F., & Gutwinski, S. (2023). Die Anwendung der Virtuellen Realität in der Behandlung psychischer Störungen. / the application of virtual reality in the treatment of mental disorders. *Der Nervenarzt, 94*(1), 27–33. https://doi.org/10.1007/s00115-022-01378-z

Turkle, S. (2011). *Alone together: Why we expect more from technology and less from each other*. Basic Books.

Turner, W. A., & Casey, L. M. (2014). Outcomes associated with virtual reality in psychological interventions: Where are we now? *Clinical Psychology Review, 34*(8), 634–644. https://doi.org/10.1016/j.cpr.2014.10.003

Usoh, M., Catena, E., Arman, S., & Slater, M. (2000). Using presence questionnaires in reality. *Presence Teleoperators & Virtual Environments, 9*(5), 497–503. https://doi.org/10.1162/105474600566989

Valmaggia, L. R., Latif, L., Kempton, M. J., & Rus-Calafell, M. (2016). Virtual reality in the psychological treatment for mental health problems: An systematic review of recent evidence. *Psychiatry Research, 236*, 189–195. https://doi.org/10.1016/j.psychres.2016.01.015

Vincelli, F., & Molinari, E. (1998). Virtual reality and imaginative techniques in clinical psychology. *Studies in Health Technology and Informatics, 58*, 67–72. https://doi.org/10.3233/978-1-60750-902-8-67

Volovik, M. G., Belova, A. N., Kuznetsov, A. N., Polevaia, A. V., Vorobyova, O. V., & Khalak, M. E. (2023). Use of virtual reality techniques to rehabilitate military veterans with post-traumatic stress disorder (review). *Sovrem Tekhnologii v Medicine, 15*(1), 74–85. https://doi.org/10.17691/stm2023.15.1.08

Wiederhold, B. K., & Bouchard, S. (2014). *Advances in virtual reality and anxiety disorders.* Springer.

Witmer, B. G., & Singer, M. J. (1998). Measuring Presencein virtual environments: A presence questionnaire. *Presence: Teleoperators & Virtual Environments, 7*(3), 225–240. https://doi.org/10.1162/105474698565686

Wolpe, J. (1973). *The practice of behavior therapy* (2nd ed.). Pergamon Press.

Zahoric, P., & Jenison, R. (1998). Presence as being-in-the-world. *Presence, Teleoperators, and Virtual Environments, 7*(1), 78–89. https://doi.org/10.1162/105474698565541

CHAPTER 2

Psychodiagnostic Assessment Through Virtual Reality

Abstract This chapter explores the transformative potential of virtual reality (VR) in psychodiagnostic assessment. It discusses the challenges and opportunities of adapting traditional paper-and-pencil tests to immersive VR environments. The chapter details the intricate process of adapting established psychodiagnostic tests to VR, addressing concerns about validity, reliability, and the preservation of essential test qualities. By navigating these complexities, it provides a comprehensive understanding of integrating traditional tests into VR. The inherent benefits and unique challenges of using VR for psychodiagnostic assessment are highlighted, emphasizing its value and versatility. VR is shown as a powerful tool that can create controlled, repeatable, and ecologically valid testing environments similar to real-world scenarios. Two case studies illustrate VR's potential in psychodiagnostic assessment. The first case study demonstrates the administration of Raven's Advanced Progressive Matrices (APM) in VR, highlighting its precision in capturing cognitive abilities. The second case study examines the administration of the Tower of London Test (TOL) in VR, showcasing the dynamic interplay between traditional assessment tools and VR's interactive capabilities. These case studies underscore VR's adaptability for assessing problem-solving and cognitive functions.

© The Author(s), under exclusive license to Springer Nature
Switzerland AG 2024
D. M. Marchioro et al., *Virtual Reality: Unlocking Emotions and
Cognitive Marvels*, Palgrave Studies in Cyberpsychology,
https://doi.org/10.1007/978-3-031-68196-7_2

Keywords Test adaptation in VR • Test validity in VR • Test reliability in VR • APM in VR • TOL in VR

2.1   ADAPTATION OF TRADITIONAL TESTS FOR VIRTUAL REALITY ENVIRONMENTS

Adapting a traditional test, that is, a test administered through paper and pencil, into a test administered in a virtual reality environment involves a series of methodological considerations on various aspects: the type of test itself, the characteristics of the subject to be examined, the reliability, and the validity and reliability of the scores obtained from the test (Parsons, 2015).

2.1.1   Type of Test

First of all, it is necessary to evaluate the type of test that is to be adapted from the traditional environment to the virtual reality environment. There are various classifications of tests, but the main one pertains to the performance required of the subject. In this context, tests are differentiated into maximum performance tests and typical performance tests (Murphy & Davidshofer, 1988; Lyman, 1991). In maximum performance tests, the subject is required to give their best, as these tests assess achieved or potential abilities. The score in these tests is determined by the level of success in correctly completing each task. These types of tests are designed to verify how well the examined subject can handle a particular situation. The scores achieved in a maximum performance test can depend on at least three factors: the subjects' innate ability, acquired ability, and motivation. These factors interact with each other, and it is not possible to assess how much of the score obtained is determined by any one of them. Performance depends both on innate potential and on how this potential has been developed and modified by life experiences (educational, social, cultural), as well as on interests and motivations. This category generally includes intelligence tests, aptitude tests, and achievement tests. Aptitude tests are used to predict how well individuals will perform; achievement tests measure their current performance, and intelligence tests assess their intellectual capacity or functioning (Pedrabissi & Santinello, 1997). In typical performance tests, all tests aim to understand a subject's preferences and habitual behaviors rather than what they can do. Unlike maximum performance tests, it is more challenging to provide clear and precise definitions for typical performance tests. There is less agreement on what they measure, leading to a significant proliferation of terms to describe them: adaptation, personality, temperament, interests, preferences, values, and so on. The tools used have various names: tests, scales, inventories, modules, indices, questionnaires, forced-choice methods, not to mention projective techniques, situational tests, and more. In these tests, the goal is not for subjects to perform as well as possible but to respond as honestly as possible. To ensure that subjects answer items honestly and not according to psychological mechanisms of desirability and adherence to social norms, some test developers, particularly those assessing personality traits, have included a subscale (known as the lie scale) to evaluate the degree of sincerity or falsehood of the respondents (Benatti & Zuin, 2017). Another issue encountered with typical performance tests is that some of their items may present some degree of ambiguity in meaning. Consequently, a subject's indecision in responding can depend on their psychological characteristics as well as the lack of clarity in a term or phrase. A subject might be perfectly sincere yet change their response to the same item over time. Such factors reduce the reliability and validity of a test. Additionally, the attitude and motivational disposition of a subject are of great importance. Those who hope to benefit from making a good impression may try to respond in a way that makes them appear better than they are, just as others might try to appear more disturbed if they believe it will be to their advantage. These behaviors can be deliberate or unconscious. This category generally includes objective personality tests, in which the range of possible responses is pre-coded and typically involves dichotomous responses (true/false) or Likert scales, and projective tests, where the range of possible responses is not pre-coded, and the subject's response involves a psychodynamic mechanism of projection (Pedrabissi & Santinello, 1997).

Another useful classification for the topics discussed is related to the time available. We thus have speed tests and power tests. In speed tests, the speed of execution of the test is a discriminating variable. There is a time limit for completing the test. Usually, speed tests consist of items with one or more correct or incorrect answers, and the significant aspect is not the difficulty level of the items but the speed in solving the proposed problems. If the time variable were removed from speed tests, most subjects would be able to correctly solve almost all items, resulting in the so-called ceiling effect. In power tests, there is no time limit, and the test administration can vary in duration, within reasonable limits. Typically, speed tests are associated with some tasks in maximum performance tests, while power tests are associated with typical performance tests (McPherson & Burns, 2008).

Finally, another classification refers to the mode of presentation of the instructions and items of the test. Thus, we have written tests, oral tests, and performance tests. In written tests, the instructions and items are presented to the subject in written form. The examiner may also read the instructions to the subject, but these are generally also presented in written form on the test protocol. Written tests are commonly referred to as paper-and-pencil tests. In oral tests, the instructions and items are presented to the subject orally. These tests are typically reserved for subjects with visual or motor impairments, or severe reading difficulties, for whom a written test would not be feasible. In performance tests, the instructions are given orally, while the items require the subject to perform a task using tools or devices (e.g., colored cubes, cards with symbols or colors, electronic or computer equipment, etc.).

Given these initial premises concerning the main classifications of tests, a fundamental question arises: is it possible to adapt all types of traditional tests to a virtual environment?

Considering the first type of classification proposed, based on the performance required of the subject (maximum performance/typical performance), it could be argued that there are no particular technical obstacles to adapting a maximum performance test from a traditional environment to a virtual reality environment. When thinking about the most commonly required tasks for a subject, such as mathematical, visuospatial, or verbal reasoning tests, there are no significant difficulties in transferring these to a virtual environment.

Considering typical performance tests, an interesting reflection emerges on the possibility of adapting projective tests to a virtual environment. Just as it is not technically problematic to display an item from an objective test and request a dichotomous or Likert scale response through virtual reality, it should not be an issue to present an ambiguous stimulus from a projective test and collect the response. However, one should question whether the projection mechanism might be influenced by the medium used, i.e., virtual reality.

Even concerning the time available (speed tests/power tests), the adaptation from a traditional environment to a virtual reality environment seems possible and feasible.

Considering the subject undergoing the test, it is essential to take into account, first and foremost, the age and the presence of any sensory or motor deficits. Age is a crucial variable as it is correlated with the perception that the subject might have of the virtual reality tool. A developing subject might be attracted to the virtual environment, associating it with a playful dimension and a gaming context. Conversely, an adult or elderly subject, especially if not accustomed to using computer technologies, might have an attitude of rejection toward the virtual environment. Therefore, beyond the technical aspect of transferring a traditional test to a virtual environment and the methodological aspects that will be analyzed later, it is fundamental that the psychologist adequately introduces the virtual reality environment to the subject. This aspect could negatively impact the subject's motivation to complete the test.

2.2   Benefits and Changes in Using VR for Psychological Assessment

The fundamental question we must ask is: why should psychological assessment take place in a virtual environment?

The virtual environment indeed constitutes a true setting with innovative characteristics compared to traditional settings. The immersive aspect, and thus the possibility of leveraging a greater number of sensory channels, is undoubtedly a significant advantage. Potentially dangerous or inaccessible virtual contexts can be recreated in total safety for the subject, or artificial scenarios that do not exist in reality can be created.

Among the benefits of VR is undoubtedly gamification, which is the use of game-derived elements and game-design techniques in non-game contexts. Studies by Thornhill-Miller and Dupont (2016) have shown that through the avatar, self-perception can change via the effect known as the Proteus effect (Yee & Bailenson, 2007): the subject subjectively experiences, through the avatar, a different approach to the problem to be faced in reality, i.e., outside the virtual environment. Through the use of the avatar, the subject can perceive different sensations and emotions, giving rise to different reasoning that can lead to alternative resolution proposals through divergent thinking.

Multiple studies suggest the utility of the VR environment for the psychological field, both in the educational sector and in evaluation, therapeutic intervention, and rehabilitation (Rizzo & Kim, 2005; Mikropoulos & Natsis, 2011).

There are, of course, negative aspects related to the virtual environment, primarily concerning: (a) the clinical conditions of the patient/client; (b) the psychologist's skills; (c) the characteristics of the room where the immersive VR practice takes place.

Regarding the clinical conditions of the patient/client, there may be difficulties related to using the equipment or states of discomfort induced by the equipment itself. User discomfort includes physical difficulties in using the tools, such as headsets and touchpads. For some people, using head-mounted displays is problematic, particularly for those with head or neck injuries or those with particular sensitivity in their eyes, as prolonged exposure to a screen a few centimeters away often causes eye strain or headaches and represents a constant issue. Additionally, there are potential states of discomfort induced by the equipment, such as cybersickness (Laviola, 2000), which refers to nausea induced by using these devices, especially in those taking medications.

Regarding the psychologist's skills, there is a technical difficulty in managing the various pieces of equipment necessary for the functioning of the virtual environment, which requires specific training.

Regarding the characteristics of the room where the immersive VR practice takes place, there may be difficulties related to space management, the correct functioning of the equipment, and the maintenance of personal safety for the user. Concerning space management, it should be noted that to ensure the normal operation of the VR system in a room, there are several technical precautions that could make it difficult to use in a psychology office. There are indeed environmental factors that interrupt the infrared detection necessary for the operation of the tools, such as climate or light reflection from windows or mirrors, which could be challenging to manage in a typical psychology office. In addition, there must be minimal maneuvering space for the user using the VR equipment, requiring a room with ample free space for use. The calibration for motion tracking of VR equipment is sensitive, and therefore, their movement must be minimized. These are not user-friendly devices as they remain bulky and require technical expertise to use. Finally, it should be remembered that the loss of spatial orientation in the real environment, resulting from the VR immersion practice, leads to a consequent issue related to the safety of the patient/client. The surrounding environment must be appropriately adapted for the safe movement of the subject and the proper functioning of the equipment. Therefore, in a psychology office context, the best solution would be to have a dedicated room—with such technical and environmental characteristics—used solely for VR immersion practices.

2.3   Considerations on Reliability and Validity in Adapting Traditional Tests to a VR Environment

After discussing the characteristics of the test that can be adapted to the virtual environment and the potential advantages and disadvantages of administering the test in VR, it becomes essential to analyze the characteristics of reliability and validity.

2.3.1   Reliability

Reliability, also known as dependability or consistency, refers to the degree of accuracy and precision of a measurement procedure. Reliability answers the question, "How accurately is the test measuring what it is supposed to measure?". A test is defined as reliable when the scores obtained by a group of subjects are consistent, stable over time, and constant after multiple administrations, in the absence of evident changes (such as psychological and physical variations of the individuals taking the test) or changes in the environment where the test takes place (Pedrabissi & Santinello, 1997).

Reliability is not a property that can be either present or absent ("all or nothing"); it can be present to various degrees, which can be calculated in different ways. In the broadest sense, test reliability indicates the extent to which differences in scores can be attributed to random measurement errors and the extent to which errors are attributable to actual differences in the characteristics being examined. Technically speaking, reliability measures of a test allow us to estimate what proportion of the total variance in scores is due to error variance (Anastasi, 2002).

The dimension of error can be calculated—for groups of subjects—through the reliability coefficient and—for individual subjects—through the standard error of measurement. Sources of error can be identified in the following phases: (a) item selection and test construction; (b) administration; (c) scoring.

Regarding item selection and test construction, the test constructor must base the test on a defined number of items extrapolated from a potentially infinite group of possible questions. Item selection is crucial for measurement accuracy. Therefore, item sampling can represent a source of measurement error. In a well-constructed test, measurement error due to item sampling will be minimal. However, it is essential to remember that a test always constitutes a sample, never the totality of an individual's knowledge or behavior.

Regarding administration, examiners constantly strive to create standardized administration conditions; however, some elements related to the application of a test can still act as error factors. Some sources of error variance can be objective conditions, due to environmental circumstances (temperature, lighting, noise, ventilation, etc.) and the materials used (dull pencils, more or less user-friendly response sheets). Another potential source of error is the subjective conditions of the person taking the test, such as health status, physical condition, prior fatigue, anxiety, level of motivation, attention and concentration, presence of emotional problems, etc. Another potential cause of error could be the presence or absence of an examiner, their physical appearance, their behavior (cold, detached, threatening), and their level of professionalism in managing the situation.

Regarding scoring, if a test's format does not include automated scoring procedures, adequate skill is required in assigning scores to various responses. Generally, most standardized tests have well-defined criteria for evaluating responses, which reduce the influence of subjective judgment. However, subjectivity in the scoring phase can pose a serious problem when using projective tests.

Transposing these considerations on possible sources of error to the virtual environment, we can observe how errors can be significantly reduced during the test administration and scoring phases. The first phase, concerning item selection and test construction, is excluded since it does not depend on the VR environment but rather on the test constructor. However, administration and scoring can have significant advantages from the VR environment.

Firstly, the administration procedures can be genuinely uniform and standardized for all subjects: the subject can read the instructions and hear a voice reading the instructions. This can also happen with paper-and-pencil tests when the administrator hands the subject the instruction sheet or activates an audio recording with the instructions. However, in the virtual environment, the entire setting can be controlled—and thus made uniform and standardized—through visual and auditory sensory channels, which are the primary ones for humans.

Regarding scoring, the administrator can achieve significantly more reliable scores because potential transcription errors by the technician are avoided: the data is directly recorded by the VR equipment. In the case of maximum performance tests with speed trials, such as latency or response duration measurements, reaction times and execution times are traditionally taken by the administrator using a stopwatch. However, this method introduces an additional source of error due to the examiner's response latency. Using VR equipment allows for accurate, automated recording of the subject's responses.

In the context of typical performance tests, using a VR environment can eliminate additional error sources, such as the transcription of responses noted on a traditional paper-and-pencil inventory. Questionnaires with several hundred responses could be inaccurately entered either by the subject—who might, for example, skip a response and move to the next line, thereby altering the responses for the entire test—or by the administrator, who could make transcription errors. In this case, a systematic data collection procedure can undoubtedly increase reliability levels (understood as measurement fidelity) by reducing possible error sources. Reducing the error source increases reliability levels.

Generally, three types of reliability measures are identified: (a) internal consistency or homogeneity; (b) stability or replicability; (c) objectivity or consistency in evaluation.

Internal consistency aims to express the degree to which all parts of the test measure the same variable under study. For example, if a test aims to measure a unitary concept, the items of that test must measure the variable in question in the same way.

Stability or replicability means that the measurement taken should not vary over time. It is important to note that psychological measurement differs from physical measurement, which can produce some issues in detection.

The last measure relates to objectivity or consistency in evaluation. A test, to be defined as such, must be standardized. This means that if the test were administered by two different evaluators to the same subjects, the results obtained should be the same. In other words, the evaluators of the test should be able to assign the same score to the same subjects.

There are different methods for calculating the reliability coefficient, varying based on the test characteristics and the conditions in which we find ourselves. Some methods require two administrations of the test (test-retest method and parallel forms method), while others require only one (split-half method and inter-rater reliability method).

Applying these theoretical aspects to the virtual environment, we observe that there can be undeniable advantages in the VR setting. The reduction of potential error sources, as previously discussed, can impact the administration and scoring procedures, thereby improving the parameters related to the objectivity and consistency of the measurement.

2.3.2   Validity

The validity of a test refers to the meaningfulness of a test score, that is, what the test score truly signifies. Validity denotes the degree of precision and accuracy with which a test measures what it is intended to measure. More specifically, it is a judgment based on the appropriateness of inferences and conclusions that can be made from the test scores (Pedrabissi & Santinello, 1997).

Validity answers the question, "What am I measuring?" The validation process of a test begins with its construction and continues over time, through the accumulation of research and clinical observations over the years. There are different forms of validity and various procedures for validating a test. Six main categories can be identified: (a) face validity; (b) ecological validity; (c) content validity; (d) construct validity; (e) criterion-related validity; (f) nomological validity.

2.3.2.1 Face Validity

Face validity refers to how convincing and relevant the tasks that the subject must perform appear. To assess this aspect of validity, one must try to adopt the perspective of the person taking the test. If the subject has a negative impression and/or finds the test useless or irrelevant to the purposes for which it is conducted, their motivation level will be low, leading to potentially skewed scores (Ercolani et al., 2002). It is crucial to remember that motivation always affects the subject's performance, and therefore, with different motivation levels, subjects' responses can change significantly.

Transitioning from the traditional environment to a virtual reality environment, it is undeniable that face validity is the most impacted aspect.

The subject, before taking the test, must wear the virtual reality equipment. At this point, it is clear that the subject's perception of virtual reality can significantly affect their motivation levels and consequently their performance in the test.

The first variable to consider is prior experience. There could be subjects who have already used VR and those trying it for the first time. In this case, there may be irrational expectations or fears, biases against VR, and stereotypes about using technology in psychology. The psychologist must adequately present the VR technology, considering the subject's background.

The second variable to consider is age, which is linked to prior experience. Children, adolescents, or young adults might be attracted to VR equipment because it is associated with gaming and entertainment. These subjects might have expectations of a recreational and entertaining experience, which will need to be appropriately adjusted. Conversely, elderly subjects who do not use electronic devices might be intimidated by VR equipment and might offer justifications related to their lack of familiarity with technology. These subjects may fear incompetence and become frustrated before starting the test or even refuse to take it. These individuals need to be guided in understanding the virtual environment and led to perceive the advantages of this mode of administration.

2.3.2.2 Ecological Validity

Ecological validity is closely related to an individual's habitual behavioral repertoire and the environment in which they typically interact. Ecological validity, therefore, expresses the degree of naturalness of the test for the individual. Imagining ecological validity along a continuum, one can identify at one end the highest degree of ecology (a test that is extremely natural for the subject, requiring habitual behaviors in a daily life context) and at the other end the highest degree of artificiality (a test entirely foreign to the subject's normal activities in an unfamiliar environment). Ecological validity must necessarily be assessed on a case-by-case basis, considering the compromises necessary to gather the required data without altering the individual's usual field of action. The higher the ecological validity of the test, the greater the ability to generalize the results to the subject's daily life.

Reflections on testing via VR are seemingly contradictory and ambivalent, as the equipment proposed for immersion in a virtual environment is undoubtedly associated with a low level of ecological validity, being artificial and far removed from the subject's normal activities. Conversely, the immersive experience via VR allows the subject to achieve high ecological validity, as various particularly realistic environments can be recreated. Therefore, we could argue that the ecological validity of VR testing is low in the initial phase of proposing the tool and preparing for immersion. However, the actual VR testing experience can present high ecological validity if realistic life environments are simulated.

2.3.2.3 Content Validity

A test is said to have good content validity when the stimulus elements it comprises (questions, graphically represented objects, etc.) produce responses that are a representative sample of the universe of content (or behavioral elements) the test aims to explore (Boncori, 1993). Content validity concerns how well the test items reflect and represent the behavior the test intends to evaluate. It involves a judgment regarding how adequately a test constitutes a representative sample within a behavioral area the test is designed to measure (Pedrabissi & Santinello, 1997).

It is believed that the VR environment can enhance the content validity of certain types of items. While it is evident that content validity depends on the test construction method, the advantages inherent in the administration mode can be significant. For example, neuropsychological tests, such as associating a food product with its corresponding price to evaluate a subject's capabilities in suspected dementia, can recreate real-life contexts: immersing the subject in a realistic environment (through virtual reality) could increase the understanding of the test, as it uses not only the verbal channel but also the visual and auditory channels, thereby potentially reducing any ambiguity of the test.

2.3.2.4 Construct Validity

Abstract variables, the result of scientists' mental constructions (hence the name construct), require measurement tools to be validated within the framework of research designs, usually complex, formulated according to the needs of scientific logic and accepted research methodology standards (Nunnally, 1978). Studies verifying construct validity aim to clarify what psychological variable the test measures: they involve logical and empirical processes and include studies on reliability, content validity, and criterion validity (Boncori, 1993). Construct validity, therefore, refers to the appropriateness of inferences made from test scores that measure a particular variable called a construct (Pedrabissi & Santinello, 1997).

The construct is a logical-hypothetical assumption adopted to predict a series of phenomena whose relationships are not observable but deducible from the adopted construct. In test construction, the problem of construct validity arises when the characteristics that define it vary or covary in such a way that they collectively constitute a functional unit suggesting reference to an identical series of otherwise unobservable phenomena (Galimberti, 1999).

To demonstrate function-related validity, it must be shown that a test has a high degree of correlation with variables it is theoretically expected to correlate with and does not significantly correlate with variables it should be unrelated to (Campbell, 1960). Convergent validation occurs when a test's construct converges with that of other tests or measures designed to assess the same or a similar construct. This method is widely used to evaluate new tests compared to existing ones or to develop parallel or shortened forms of the same test. In these cases, to ensure that both tests measure the same construct, a correlation coefficient greater than 0.80 should be obtained. Discriminant validation is similar in procedure to the previous one, but low correlations with other variables are sought, with which, from a theoretical construct perspective, there should be no influence. Discriminant validation is especially important for personality tests, where irrelevant variables can variously affect scores (Anastasi, 2002).

Administering the same test in traditional and virtual environments, thus correlating the subject's performance in two identical tests that differ only in the administration environment, represents a convergent validation process. As will be seen in the case study on Raven's Advanced Progressive Matrices (APM) (Marchioro et al., 2023), it is possible to validate the traditional tool, transposed into a virtual environment, precisely through a convergent validation methodology.

2.3.2.5 *Criterion-Related Validity*

If a test result is to be used to draw inferences about some type of behavior (e.g., academic success or failure) or psychological states (e.g., reduction in a dysphoric state following psychotherapy), an empirical verification must be conducted where the subject's behaviors or states, assumed as criteria, are compared with their test responses. In these cases, we talk about validity concerning an external criterion. If the behavior or psychological state assumed as a criterion is measured simultaneously or almost simultaneously with the test administration, it is referred to as concurrent validity; if the time interval between the two measurements is such that some change might have occurred (e.g., due to learning, therapy, etc.), it is referred to as predictive validity (Boncori, 1993). Criterion-related validity is the characteristic of a test that allows us to understand how adequately the result can be used to predict a subject's future performance in a particular activity or another test (Pedrabissi & Santinello, 1997).

In essence, this evaluates how test scores are connected to other measures, and this correlation level expresses the relative validity level. The criterion should be a credible variable, adequate to document what the test really measures, and itself measured validly and reliably (Boncori, 1993). The criterion can be a standard, a set of elements to base a judgment or decision on, another test score, a psychiatric diagnosis, a time measure, etc. (Pedrabissi & Santinello, 1997).

The important thing is that the criterion is a direct, but external and independent, measure of the same variable or trait that the test aims to measure. In practice, the criterion is another measure of what the test wants to measure, but it is a measure made with different procedures from those of the test, constituting an external and independent reference term. It is crucial that the criterion data are not contaminated (if contaminated, the data are no longer external and independent).

Even for this type of validity, the virtual environment can provide undeniable advantages. The possibility of immersive experience can expand the range of possible criteria to consider in the validation process. For example, in cases of anxiety and phobia issues, the ability to bring the subject into contact with potential phobic stimuli in a controlled environment can be highly beneficial.

2.3.2.6 *Nomological Validity*

Nomological validity is the most general, abstract, and simultaneously the most challenging form of validity to verify. It can be defined as the degree to which the set of hypotheses related to the measurement of the construct under examination are verified within a broader conceptual framework that includes other constructs with which the measurement of the construct of interest maintains theoretically justified relationships. From this concept, it can be deduced that nomological validity encompasses all other forms of validity. Its more complex meaning, however, can be better explicated and understood if conceptualized as the large-scale extension of criterion validity. Unlike criterion validity, which is based on the association between two variables at a time, nomological validity is summarized in a complex of associations between the construct and multiple criteria (Ercolani et al., 2002).

2.4 Case Study 1: Administering Raven's Advanced Progressive Matrices (APM) in a VR Setting

Raven's Progressive Matrices represent a general cognitive test used to examine mental abilities that can be applied to anyone, regardless of cultural level and age group. It is a good tool for assessing the ability to analyze, construct, and integrate a series of concepts. Raven's Progressive Matrices primarily concern the visuospatial aspect and the ability to analyze abstract figures based on similarity, difference, numerical progression, size, and more. They are among the most widely used tools for measuring fluid intelligence and require analyzing, constructing, and integrating a series of concepts directly without resorting to subscales or summation of secondary factors. They consist of an abstract drawing with a missing piece in the bottom right corner and a series of alternatives that complete the drawing, which the subject must indicate or name.

There are three different versions of Raven's Progressive Matrices based on age and education level: (a) Coloured Progressive Matrices (CPM); (b) Standard Progressive Matrices (SPM); (c) Advanced Progressive Matrices (APM).

The Coloured Progressive Matrices (CPM) (Raven, 1947) are a simplified version of the original test, suitable for examining children, the elderly, and pathological subjects such as aphasics, those with mental retardation, or cerebral palsy. The CPM consists of 36 items divided into three series (A, Ab, B), each with 12 items. In Italy, the first publication dates back to 1984, while the current adaptation was conducted on 4877 children aged between 3 years and 11 years and 6 months and 386 adults and elderly individuals aged between 55 and 93 years (Belacchi et al., 2008). In Fig. 2.1 you can see the first item of CPM.

The Standard Progressive Matrices (SPM) (Raven, 1941) represent the original version of the test aimed at evaluating normally functioning adults. The SPM consists of 60 items divided into five series (A, B, C, D, E), each with 12 items, in increasing order of difficulty. In Italy, the first publication dates back to 1954 (Raven, 1954), while the current adaptation was conducted on 845 non-clinical subjects (443 males and 402 females) and 170 clinical subjects (95 males and 75 females) (Raven, 2008). In Fig. 2.2 you can see the first item of SPM.

The Advanced Progressive Matrices (APM) (Raven, 1948) represent the most challenging version of the SPM, designed for application on adults with above-average abilities. The APM consists of 48 items divided

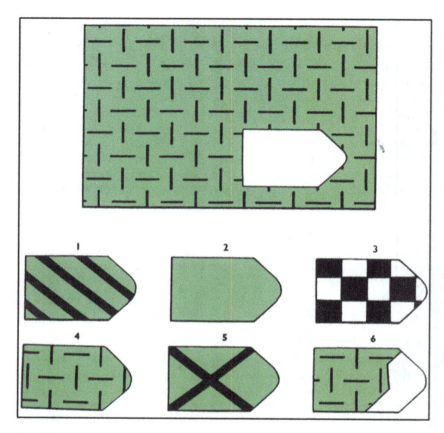

Fig. 2.1   Item 1 Serie A CPM. (Reproduced from Raven, 1947)

into two series: Series I with 12 items and Series II with 36 items. In Italy, the first publication dates back to 1969 (Raven, 1969), while the current adaptation was conducted on 960 students in the fourth and fifth years of secondary school, 679 university students, and 127 adults undergoing selection (Di Fabio & Clarotti, 2007). In Fig. 2.3 you can see the first item of APM.

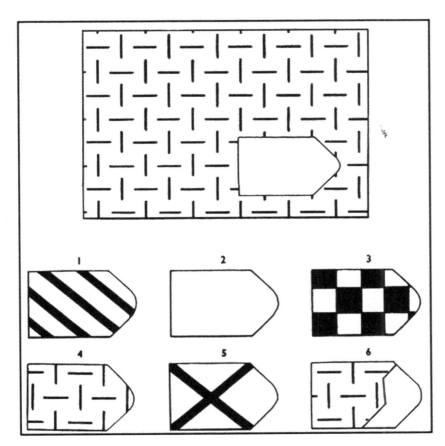

Fig. 2.2   Item 1 Serie A SPM. (Reproduced from Raven, 1941)

2.4.1   Advanced Progressive Matrices (APM) in a Virtual Environment

In the study conducted by Marchioro et al. (2023), the 48 items of the APM were divided into two blocks of 24 items each. The division of the items, using the odd-even method, ensured a balanced difficulty level between each block of items. Participants were randomly assigned to two groups, and in each group, the administration of the two blocks of items was alternated as follows. In Group 1, participants started with 24 items

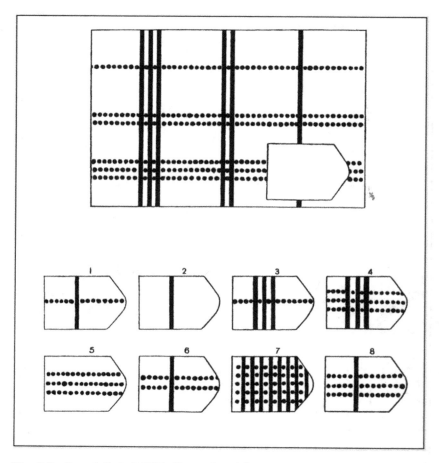

Fig. 2.3   Item 1 Serie I SPM. (Reproduced from Raven, 1948)

administered in a real environment using the traditional method, then completed the second half of the items using virtual reality. In Group 2, participants started with the first 24 items in a virtual environment and then completed the second half of the items using the traditional method. Each participant thus experienced both the real and virtual versions of the APM test, allowing direct within-subject comparisons of problem-solving performance and reducing the influence of individual differences. Randomization of the administration order further enhanced the study's validity by minimizing potential sequence effects.

To ensure consistency, the virtual environment was designed to be as realistic as possible to the real environment in terms of aesthetics and acoustics. Through the equipment, data on participants' problem-solving performance were collected, including response time, accuracy, and solution strategies in both real and virtual environments. These parameters provided a comprehensive assessment of cognitive performance. Among the variables considered, participants' prior experience with virtual reality technology (VRT) was also examined as a potential influencing factor.

The analysis conducted to evaluate the differences between real and virtual environments revealed no significant difference in problem-solving performance between the environments (real vs. virtual). However, significant differences emerged in the response time of the subjects.

The analysis, conducted using a mixed-model analysis of variance (ANOVA), revealed no significant differences in problem-solving performance between the real and virtual environments, regardless of prior experience with VR. The main effects for the "Group" variable and the "VRT previous experience" variable were not significant, indicating that the levels of these variables were similar for both environments, virtual and real. Furthermore, the within-subject factor and interaction effects were not significant, suggesting no significant differences in problem-solving performance between the two environments (Table 2.1).

To provide a more detailed analysis, a multivariate analysis of variance (MANOVA) was conducted to examine if there were significant differences in the mean performance time between virtual and real environments. However, the main effects for the variables "Group," "Gender," and "VRT previous experience" were not significant, indicating that there were no significant differences in performance time between these variables (Table 2.2).

For a more comprehensive analysis, a two-tailed paired-samples t-test was conducted within each group to determine if there were significant differences between the scores obtained in the two blocks of 24 items presented in different orders. In Groups 1 and 2, the difference in means was not significantly different from zero (Fig. 2.4).

Regarding response time, significant differences were observed across the entire sample. Participants' response times were significantly shorter in the virtual environment compared to the real environment, suggesting that the virtual environment facilitated concentration ($t = -5.05$, $p < 0.001$, Cohen's $d = -0.955$).

Table 2.1 Mixed Model ANOVA Results (Reproduced from Marchioro et al., 2023)

Source	df	SS	MS	F	p	η_p^2
Between-Subjects						
Group	1	2.62	2.62	0.09	0.770	0.003
VRT previous experience	1	1.68	1.68	0.06	0.815	0.002
Residuals	25	751.70	30.07			
Within-Subjects						
Within Factor	1	2.75	2.75	0.66	0.424	0.03
Group: Within.Factor	1	4.48	4.48	1.08	0.310	0.04
VRT previous experience: Within.Factor	1	0.002	0.002	0.00	0.984	0.00002
Residuals	25	104.08	4.16			

Table 2.2 Mixed Model ANOVA Results (Reproduced from Marchioro et al., 2023)

Variable	Pillai	F	df	Residual df	p	η_p^2
Group	0.09	1.16	2	23	0.330	0.09
Gender	0.01	0.10	2	23	0.909	0.01
VRT previous experience	0.09	1.11	2	23	0.345	0.09

Posthocs. Since there were no significant predictors, additional testing was not performed

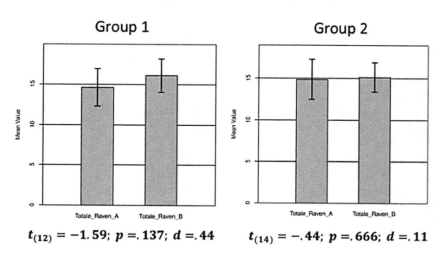

$t_{(12)} = -1.59; \ p = .137; \ d = .44$ $t_{(14)} = -.44; \ p = .666; \ d = .11$

Fig. 2.4 Two-tailed paired-samples *t* test. (Reproduced from Marchioro et al., 2023)

2.5   CASE STUDY 2: TOWER OF LONDON TEST (TOL) IN A VR SETTING

The Tower of London (TOL) test is one of the most widely used tests for studying and measuring planning and task-monitoring abilities in both clinical and research activities. To solve many problems, it is necessary to anticipate and keep in mind the consequences of one action on others. This interdependence of elements in a complex problem is a characteristic of many everyday situations. TOL aims to probe executive functions (Eslinger, 1996), specifically planning and problem-solving.

Shallice (1982) identifies the prefrontal cortex as responsible for predicting one's actions and recognizing them as appropriate or not for the purpose. If a subject is presented with problems they have already learned to tackle, they will activate automatic resolution schemes managed by the selective system for similar problems. When facing new problems or dilemmas that require alternative solutions to those already schematized, the Supervisory Attentional System is activated. This system requires effort and energy but ensures reliability by constantly monitoring the entire resolution process, regulating all forms of bodily and mental activity.

The Italian edition of TOL (Sannio Fancello et al., 2006) was standardized on a sample of 1772 children aged 4 to 13 years.

In performing the tasks, three operations are specifically required: (a) formulation of a general plan; (b) identification of sub-goals to organize within a sequence of movements; (c) maintenance of sub-goals and the general plan in working memory (WM).

The material consists of three pegs (1, 2, 3) of different lengths mounted on a rectangular structure and three balls of different colors: red (R), green (G), blue (B), which need to be placed on the pegs.

TOL comprises a series of 12 items of gradual difficulty, depending on the number of moves the subject must make to reach the configuration shown by the examiner. The test always starts from a predefined base position (Fig. 2.5).

The administration procedure includes a series of measurements:

Movement Initiation Time: Determined by the response latency between the start of each trial and the touching of the first ball.

Motor Execution Time: Determined by the duration of completing the individual trial, i.e., the time between touching the first ball and completing the sequence of moves necessary to solve the problem.

Fig. 2.5 TOL
predefined base position

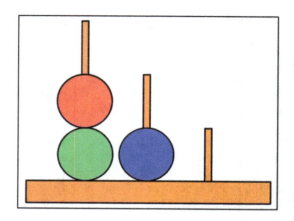

Total Motor Execution Time: Sum of the two previous variables. Dividing the total motor execution time by the number of moves provides an estimate of the average movement time per move.

Initial Planning Time: The time between the presentation of the problem and the touching of the first ball, minus the corresponding movement initiation time.

Subsequent Planning Time: The time between touching the first ball and completing the sequence of moves necessary to solve the problem, minus the corresponding motor execution time. Dividing the scores by the number of moves provides an estimate of the average planning time per move.

The psychologist verbally presents the following instructions to the subject: "Here are three sticks of different lengths and three balls of different colors. You need to arrange the balls according to the configurations I will show you. The figure on the sheet shows one such arrangement. Now you need to copy this figure to ensure you have no problems recognizing the colors. I will now show you another figure and ask you to change the balls from this configuration here to a different one, but there are rules to follow: you can move only one ball at a time; you can move from one stick to another only. Thus, you cannot place a ball on the table or hold more than one ball at a time; you can place only one ball on the short stick, two on the medium stick, and three on the long stick. If you follow this rule, the balls will not fall off the stick. I will tell you each time how many moves are necessary to solve the problem."

Regarding administration, it is advisable to administer TOL in a single session. It is important to always start with the sample problem and administer all 12 trials (items) unless the subject shows they cannot solve the sample problem (Fig. 2.6).

It's important to note any rule violations, such as moving more than one ball at a time, holding a ball or placing it on the table while moving another, or placing too many balls on one stick. Each rule violation is assigned 1 point, and the total score is the sum of violations across all items. This variable provides insights into the subject's ability to understand and remember the rules presented for task execution.

The total accuracy score is calculated using the following scoring rules: each problem is considered solved only if completed in the specified number of moves; if solved on the first attempt, 3 points are awarded; if solved on the second attempt, 2 points; if solved on the third attempt, 1 point; in all other cases, 0 points are awarded. This parameter measures the subject's planning ability. With 12 items, the maximum achievable score is 36 points.

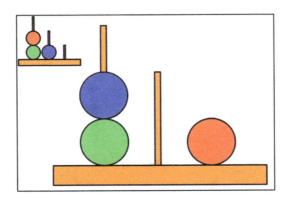

Fig. 2.6 TOL initial example. TOL predefined base position is shown in the top left corner. Recording Responses in TOL. The psychologist notes the subject's moves during each trial using the following alphanumeric coding: R = red ball; B = blue ball; G = green ball; 1 = short stick; 2 = medium stick; 3 = long stick. A move is defined as when a ball is removed from one stick and placed on another or repositioned on the same stick

Decision Time: The time interval between the start signal given by the examiner and the first movement made by the subject when beginning to solve the task. This time is summed across all 12 trials.

Execution Time: The duration of the movement until the completion of that particular attempt. Incidental movements on the apparatus are excluded. This time is summed across all 12 trials.

Total Time: The sum of decision time and execution time for all trials performed by the subject in the 12 trials.

2.5.1   Tower of London (TOL) in a Virtual Environment

The experiment (Marchioro & Benatti, 2022) was conducted using a virtual reality system running a specially created simulation software. The system consisted of a computer with the necessary technical specifications and graphics card to support the high-quality simulation, and a virtual reality set comprising a Head Mounted Display (HMD) (Fig. 2.7) and two touchpad input devices (Fig. 2.8) used to track hand movements.

Hardware Specifications:

Computer: Dell Alienware model with a latest-generation Intel i7 processor, 32GB RAM, SSD hard drive, three USB 3.0 ports, and an NVIDIA GeForce 1080 graphics card.

Virtual Reality Set: Oculus Rift, providing six degrees of freedom motion tracking, a video resolution of 2160 by 1200 pixels (1080 by 1200 pixels per eye), a field of view greater than 90°, 3D audio (allowing the

Fig. 2.7 Oculus Head Mounted Display. (Reproduced from Marchioro & Benatti, 2022)

Fig. 2.8   Oculus Touchpads. (Reproduced from Marchioro & Benatti, 2022)

determination of the three-dimensional origin of a sound within the simulation), and two Oculus touchpads.

The virtual reality simulation software was developed using Unreal Engine, a high-quality game development environment typically used for AAA games. The simulation software was built from scratch following the specific requirements and constraints of the TOL test. Programming was done using Unreal Engine's proprietary visual system called Blueprint and the C++ language. Materials, models, sounds, and music were sourced from various online libraries, while architectural models and scenario compositions were created in-house. The creation and testing of the simulation for release took approximately two months.

Interaction Management:

Users were trained to grasp the spheres and place them on the pegs in a TOL simulation (Fig. 2.9).

The TOL scenario (Fig. 2.10) places the user in a small room with the movement system disabled. In front of them, they can see the set of pegs and spheres needed for the test and a panel displaying instructions and results.

The operator overseeing the tests could sequentially initiate all the items defined for that specific phase of the simulation. All test results were exported to an Excel file. In this setup, the user is immersed in a realistic environment that mirrors the traditional TOL test setting, allowing for accurate data collection on problem-solving performance.

Fig. 2.9 VR TOL Training Map. (Reproduced from Marchioro & Benatti, 2022)

Fig. 2.10 VR TOL Scenario. (Reproduced from Marchioro & Benatti, 2022)

REFERENCES

Anastasi, A. (2002). *I test psicologici*. Milano.

Belacchi, C., Scalisi, T. G., Cannoni, E., & Cornoldi, C. (2008). *Manuale CPM Coloured Progressive Matrices, Standardizzazione italiana*. Giunti O.S.

Benatti, F., & Zuin, A. (2017). *Elaborazione e requisiti delle prove psicodiagnostiche – Test 1* (Seconda edizione). Libreriauniversitaria.it Edizioni.

Boncori, L. (1993). *Teoria e tecniche dei test*. Bollati Boringhieri.

Campbell, D. T. (1960). Recommendations for APA test standards regarding construct, trait, or discriminant validity. *American Psychologist, 15*(8), 546–553. https://doi.org/10.1037/h0048255

Di Fabio, A., & Clarotti, C. A. (2007). *Le Matrici Progressive di Raven: CPM, SPM, APM. Organizzazioni Speciali*. Giunti O.S.

Ercolani, A. P., Areni, A., & Mannetti, L. (2002). *Metodologia della ricerca psicologica*. Il Mulino.

Eslinger, P. J. (1996). Conceptualizing, describing, and measuring components of executive function: A summary. In G. R. Lyon & N. A. Krasnegor (Eds.), *Attention, memory, and executive function* (pp. 367–395). Paul H Brookes Publishing.

Galimberti, U. (Ed.). (1999). *Enciclopedia di Psicologia*. Torino.

Laviola, J. J. (2000). A discussion of cybersickness in virtual environments. *ACM SIGCHI Bulletin, 32*(1), 47–56. https://doi.org/10.1145/333329.333344

Lyman, H. B. (1991). *Test scores and what they mean*. Allyn & Bacon.

Marchioro, D., Arguero Fonseca, A., Benatti, F., & Bounous, M. (2023). *Effectiveness of images with high-potential restorative in virtual reality to reduce acute cognitive fatigue* in undergraduate students. Oral presentation presented at the *18th European Congress of Psychology (ECP) "Psychology: Uniting communities for a sustainable world"*.

Marchioro, D., & Benatti, F. (2022). *Processes of planning in a virtual reality experience: link between arousal and problem solving*. Oral presentation presented at the *17th European Congress of Psychology (ECP) "Psychology as the Hub Science: Opportunities and Responsibility"*.

McPherson, K., & Burns, N. R. (2008). Assessing the validity of computer-game-like tests of processing speed and working memory. *Behavior Research Methods, 40*(4), 969–981. https://doi.org/10.3758/BRM.40.4.969

Mikropoulos, T. A., & Natsis, A. (2011). Educational virtual environments: A ten-year review of empirical research (1999–2009). *Computers & Education, 56*(3), 769–780. https://doi.org/10.1016/j.compedu.2010.10.020

Murphy, K. R., & Davidshofer, C. O. (1988). *Psychological testing: Principles and applications*. Prentice Hall.

Nunnally, J. C. (1978). *Psychometric theory* (2nd ed.). McGraw-Hill.

Parsons, T. D. (2015). Virtual reality for enhanced ecological validity and experimental control in the clinical, affective and social neurosciences. *Frontiers in Human Neuroscience, 9*, 660. https://doi.org/10.3389/fnhum.2015.00660

Pedrabissi, L., & Santinello, M. (1997). *I test psicologici. Teorie e tecniche*. Il Mulino.

Raven, J. C. (1941). *Standard progressive matrices: Sets A, B, C, D & E*. H. K. Lewis.

Raven, J. C. (1947). *Coloured progressive matrices: Sets A, Ab, B*. H. K. Lewis.

Raven, J. C. (1948). *Advanced progressive matrices: Set I & II*. H. K. Lewis.

Raven, J. C. (1954). *Manual for the Raven's progressive matrices and vocabulary scales*. H. K. Lewis.

Raven, J. C. (1969). *Advanced progressive matrices*. H. K. Lewis.

Raven, J. C. (2008). *Standardizzazione italiana delle Matrici Progressive di Raven: Manuale SPM Standard Progressive Matrices*. Giunti O.S.

Rizzo, A. S., & Kim, G. J. (2005). A SWOT analysis of the field of virtual reality rehabilitation and therapy. *Presence: Teleoperators and Virtual Environments, 14*(2), 119–146. https://doi.org/10.1162/1054746053967094

Sannio Fancello, G., Vio, C., & Cianchetti, C. (2006). *Tower of London test: Manuale*. Giunti O.S.

Shallice, T. (1982). Specific impairments of planning. *Philosophical Transactions of the Royal Society of London. B, Biological Sciences, 298*(1089), 199–209. https://doi.org/10.1098/rstb.1982.0082

Thornhill-Miller, B., & Dupont, J. M. (2016). Virtual reality and the enhancement of creativity and innovation: Under recognized potential among converging technologies? *Journal of Cognitive Education and Psychology, 15*(1), 102–121. https://doi.org/10.1891/1945-8959.15.1.102

Yee, N., & Bailenson, J. N. (2007). The Proteus effect: The effect of transformed self-representation on behavior. *Human Communication Research, 33*(3), 271–290. https://doi.org/10.1111/j.1468-2958.2007.00299.x

CHAPTER 3

Immersiveness and Its Effect in Cognitive Functions

Abstract In the second chapter of our exploration into the marriage of virtual reality and psychology, we delve into the captivating realm of immersiveness and its profound impact on cognitive functions. This chapter explores how the immersive qualities of virtual reality can captivate and influence the human mind, driving cognitive engagement within virtual environments. By eliciting deep and authentic emotional responses, VR environments make the study of cognitive functions and emotions more ecologically valid. A key focus of the chapter is the investigation of problem-solving within virtual reality. It examines the cognitive processes at play when individuals are immersed in digital worlds, addressing how VR affects problem-solving abilities, decision-making, and critical thinking. Central to this discussion is a compelling case study titled "Processes of Planning in a Virtual Reality Experience: The Link Between Arousal and Problem Solving." This research explores the connection between emotional arousal and problem-solving skills in a VR setting, highlighting the interplay between emotional states and cognitive abilities. As the chapter progresses, it becomes evident that immersiveness in VR has the potential to transform not only the study of cognitive functions but also practical applications in education and therapy. The intersection of emotions, cognitive processes, and VR's immersive qualities offers vast opportunities for researchers, practitioners, and enthusiasts.

© The Author(s), under exclusive license to Springer Nature   61
Switzerland AG 2024
D. M. Marchioro et al., *Virtual Reality: Unlocking Emotions and Cognitive Marvels*, Palgrave Studies in Cyberpsychology,
https://doi.org/10.1007/978-3-031-68196-7_3

Keywords Virtual reality • Immersiveness • Cognitive functions •
Arousal • Problem-solving • Emotions

3.1   IMMERSION AND PRESENCE IN VIRTUAL
REALITY ENVIRONMENTS

As elaborated in the preface, the concept of "immersion" evokes the
Platonic myth of the cave, where individuals confuse the shadows cast on
the wall with reality itself. In a sense, virtual reality (VR) creates a sort of
modern "cave," where the shadows are replaced by digital simulations.
However, unlike Plato's myth, the aim of VR should be to enlighten and
expand our understanding of psychological and therapeutic reality, rather
than deceive us.

Immersion can be defined as the psychological state in which one perceives being enveloped and included in an environment that provides a
continuous stream of stimuli and experiences. According to Szabó and
Gilányi (2020), there are two main schools of thought providing definitions of immersion in VR, and their definitions significantly differ. One
school, associated with Witmer and Singer (1998), considers immersion as
a psychological state of the user. The other school, based on the works of
Slater (Slater et al., 1996; Slater, 1999; Slater, 2003), views immersion as
an objective characteristic of a VR system.

Witmer and Singer (1998) delve into the concept of "presence" in virtual reality contexts. They define presence as the subjective experience of
being in a specific place or environment, even when physically situated
elsewhere. This sense of presence is influenced by the interaction between
sensory stimulation and environmental factors that encourage immersion
and engagement, as well as by the individual's internal tendencies to
actively engage.

According to Witmer and Singer (1998), immersion is a psychological
state in which the individual perceives themselves to be enveloped by,
included in, and interacting with an environment that provides a continuous flow of stimuli and experiences. Thus, immersion is deeply linked to
the quality of the VR environment and the modes of interaction it allows.
For example, a VR environment that effectively isolates the user from the
external physical environment, providing natural interaction modes and a
realistic perception of movement, tends to significantly increase the sense
of immersion and, consequently, of presence.

In technological terms, immersion is manifested through the use of devices such as virtual reality headsets, which enhance sensory isolation from the external environment, allowing for a more engaging and immersive experience. However, immersion is not just an objective feature of the VR environment but also a subjective experience of the user, which depends on their ability to focus attention and engage with the virtual environment.

Witmer and Singer (1998) also emphasize the importance of the interaction between factors of immersion and involvement, indicating that both are necessary for a full experience of presence in VR. They have developed the Presence Questionnaire (PQ) to measure presence in VR environments, based on these factors and how they influence the user's experience.

Their research opens significant reflections on how VR environments can be optimized to improve the sense of presence, through intelligent manipulation of immersion and involvement factors, which are crucial for the effectiveness of such technologies in applications ranging from entertainment to education and therapy.

The virtual environment that effectively isolates users from the physical environment, providing natural interaction modes and perception of movement, increases the degree of immersion and, therefore, presence. The ability of VR to create immersive experiences has enabled psychologists to study cognitive and emotional responses in ways previously unreachable with traditional methods.

Slater (1999) discusses the concept of immersion in virtual reality environments, emphasizing its fundamental role in inducing a compelling experience for the user. According to Slater (1999), immersion is primarily defined by the technological characteristics and configuration of the VR system. These include the quality and consistency of visual representations, audio and tactile feedback, and the system's ability to respond realistically and promptly to user inputs. Slater (1999) argues that these technological aspects create an environment that can effectively "deceive" the senses, making users feel as if they are physically present within the virtual environment. This sensation of presence is crucial for VR applications, as a higher level of immersion can significantly enhance the effectiveness of the experience, whether it is aimed at entertainment, training, or therapy.

The author distinguishes between the concept of "immersion," understood as an objective quality of VR technology, and "presence," which is the user's subjective experience of being in a place different from their actual physical environment. According to Slater (1999), the key to an effective VR experience lies in the ability to maximize immersion through optimal use of available technologies, thereby promoting a strong sense of presence in the virtual context. This approach offers a comprehensive view of the importance of hardware and software in shaping the virtual reality experience, highlighting the dynamic interaction between the user and the technology that defines the very experience of immersion.

Practical applications, already anticipated in the introduction, show how the use of VR in the treatment of anxiety disorders—using safe and controllable environments—can facilitate exposure therapy, overcoming the logistical and ethical constraints of real environments (Botella et al., 2015).

Recent studies emphasize how the immersive nature of VR can intensify emotional involvement and cognitive presence, fundamental elements in therapeutic and experimental contexts (Turner & Casey, 2014). This phenomenological approach places the individual's lived experience at the center, essential for understanding cognitive and emotional dynamics.

An immersive virtual environment is a computer-generated environment that evokes the user's sense of presence. It is defined as an aesthetic perception that invites total immersion in the virtual space and a voluntary suspension of disbelief (Nechvatal, 2009).

Furthermore, there are two other elements dependent on technology development that influence the sense of presence: vividness and interactivity. The first dimension is understood as the production of a mediated environment rich in sensations, the second is defined as the user's ability to interact with the environment and modify its shape or alter events through interaction with it (Garrett et al., 2017).

The ability of VR to simulate complex environments and elicit authentic emotional responses invites us to reflect more deeply on the nature of our perception and cognitive experiences. VR offers an unprecedented laboratory to observe how the human brain interacts with realities that, although artificial, provoke reactions very similar to those that would occur in a non-virtual context. This leads us to consider the mind not only as a biological entity but also as a dynamic and flexible construct, capable of adapting to and being influenced by virtual environments.

Currently, there are various types of VR devices that allow for qualitatively different experiences. The most well-known and used are the aforementioned Oculus Rift and HTC Vive. Although these two systems are the most widespread and thus the most supported by development software libraries, there are other VR systems such as PlayStation VR, Razer OSVR, Pimax 4 K, Dell Visor, HP Reverb, Lenovo Explorer, Samsung Odyssey, and Microsoft HoloLens. There are also less performant systems from the point of view of visors and sensors and, consequently, less expensive, with which it is still possible to have a VR experience, albeit much less immersive, such as Samsung Gear VR, Google Cardboard, or Oculus GO.

The understanding and application of theories of immersion in virtual reality promise to revolutionize therapeutic practices and research in the field of psychology, offering new possibilities for more effective treatments and more in-depth studies of psychological and behavioral dynamics.

3.2   PROBLEM-SOLVING

What types of skills are necessary for problem-solving? Is there a general capability to address problematic situations, which can then be refined to tackle specific cases, or does each type of problem require unique skills? Problem-solving comes into play when it is necessary to overcome obstacles to achieve a particular goal; indeed, we would not be facing a problem if the solution were readily retrievable from memory (Magro & Muffolini, 2011). Hayes (1989) and Bransford and Stein (1993) have identified various operations in the problem-solving cycle that include: (a) problem identification; (b) definition of the problem; (c) strategy formulation; (d) organization of information; (e) allocation of resources; (f) monitoring; (g) evaluation.

Problem Identification. Recognizing that a particular situation poses a problem is not always straightforward, as it can be challenging to identify a goal, or the path to achieving it may be fraught with difficulties, or the envisioned solution may not work. For instance, if the problem involves preparing a computer-based graphical presentation, one must first be acquainted with the subject of the presentation.

Problem Definition and Representation. The next step is to define and represent the problem adequately; if it is defined or represented inaccurately or imprecisely, it becomes more challenging to solve. Continuing with the example, it is crucial to define the topic carefully and accurately to facilitate the search for documentation and to outline the general strategy to be employed.

Formulation of a Strategy. Once the problem is defined, a resolution strategy must be planned. It might be beneficial to employ an analytical process, breaking down the complex problem, or a synthetic strategy, which involves combining different elements in a useful manner (analyzing and identifying various aspects).

Organization of Information. The available information is then organized within a framework that allows the chosen strategy to be applied most effectively (creating an index of parts to divide the presentation into).

Resource Allocation. It is necessary to address the problem related to the availability of limited resources: time, money, space, tools, etc. Some problems require substantial time and resources, while others may need less. Research on samples of more or less experienced students has shown that experts in problem-solving tend to allocate more mental resources to overall planning focused on the big picture, whereas novices spend more time on local planning, focused on details (deciding which program to use among those available based on the time and skills available).

Monitoring. Effective problem-solvers, after choosing a particular solution path, frequently check the steps they are actually taking to reach their goal (verifying the consistency of the individual images prepared, the use of the same graphic criteria, the timing of the images' display, etc.).

Evaluation. In addition to monitoring the problem during the solution process, it is necessary to evaluate the achieved solution once the process is completed; some evaluations can occur quickly, while others might only be possible later or much after the solution is implemented (simulating the presentation to the client or other people to see if the goal has been achieved).

According to some scholars (Greeno, 1978; Reed, 1988), there are general abilities that, however, fully express themselves especially in certain types of problems. Cavallin (2015) defines problem-solving as a thought process that, starting from a given and perceived unsatisfactory condition, leads to the identification and achievement of a desired situation considered an improvement of the existing one. It allows for bridging a gap between reality and desire.

Gestalt psychologists believe that the mechanism underlying the problem-solving process is insight, which is the sudden and unpredictable appearance of the resolution without having the slightest idea of what happened, commonly depicted in popular iconography by a light bulb turning on, as a form of sudden enlightenment (Köhler, 1917).

Often, insight involves the re-conceptualization of a problem or strategy in an original way, or the combination of already known relevant information with new information, to arrive at the solution through an innovative perspective.

Duncker (1945) observed that problem-solving is usually not achieved in a single operation, but involves a succession of insights; moreover, previous experiences influence the solution processes.

These insights are found in Wertheimer (1959), who discusses reproductive or structurally blind thinking, used mechanically and inappropriately for the specific situation, based solely on the application of knowledge; in contrast, he describes productive thinking as the method that identifies the general principles and fundamental requirements of the problem, upon which the search for a solution is based.

Davidson and Sternberg (1984) believe that insight requires three types of selective skills for problem-solving: encoding (identifying what information is important among all available), combining (discovering a new arrangement of seemingly disparate elements), and comparison (a more original comparison than usual methods).

Metcalfe and Wiebe (1987) distinguished two modes of problem-solving: in the first, not accompanied by insight, the subject progressively builds the solution and experiences a feeling of warmth, increasing as they approach the resolution; in this case, the sense of knowing allows a certain confidence in providing the solution without the help of insight. In the second mode, where the solution is accompanied by a sudden insight, there is no such progression, and the subject significantly raises the sensation of enthusiasm, which has remained constant at a certain level until then.

Subsequent research has shown that progressive solutions by insight can arise from gradual unconscious processes (Bowers et al., 1990).

Insight leads to considering elements of the problem previously ignored and manifests thanks to a temporary distancing from the problem. Insight is not an intentionally reproducible process as the individual can only try to foster it by creating environmental conditions conducive to its generation, but cannot intervene directly in it (Cavallin, 2015).

3.3   CREATIVITY

Creativity is regarded as our most significant individual and collective capacity and is sometimes even identified as the characteristic that makes us human (Thornhill-Miller & Dupont, 2016). Creativity is such an overused term that it often remains poorly defined, vague, and certainly not clarified by the numerous definitions provided by scholars over the years. Generally, we can consider creativity as a thought activity that directs and sustains human energy toward achieving new, useful, comprehensible, coherent, and environmentally compatible results (Cavallin, 2015).

However, creativity is not always easy to summon in problem-solving: anyone can feel stuck for various reasons—fatigue, a bad day, distracting thoughts—there are countless reasons behind an inactive mind (Cavallin, 2015). Over the years, solutions have been sought to overcome these situations: studies have suggested a more informal work environment, replacing traditional chairs with sitting balls to ease the body (almost as a way to not limit it, thus not limiting the mind), walls in creative offices have been painted with stimulating colors (Thornhill-Miller & Dupont, 2016), and more flexible working hours have been introduced (Anderson, 1994). Dul and Ceylan (2011) have also provided evidence that specific aspects of the physical environment correlate with increased creativity in work settings. Some of these environmental elements, such as the presence of windows, calming or stimulating colors, and plants or other natural components, can influence creativity indirectly through processes of arousal and emotionality previously discussed.

From the study on the relationship between bodily experience and creativity, it has been reported that an arm capable of moving fluidly can facilitate creative cognition (Slepian & Ambady, 2012), just as people are more likely to solve problems if they have an open space in which they can freely walk (Leung et al., 2012). It has also been demonstrated that, in general, compared to sitting, walking is an excellent activity for generating ideas with greater fluidity and originality (Oppezzo & Schwartz, 2014). Moreover, there are environmental elements that appear to contribute to creativity by functioning at a more cognitive level, such as problem-solving visualized through stereotypical metaphors, for example, the presence of a lit light bulb (Slepian et al., 2010). A study explored whether the embodiment of the metaphor breaking the rules could influence creative performance using VR technology to facilitate this test by allowing participants to destroy walls. The results revealed positive effects of wall-breaking on creative performance, with greater originality, fluidity, flexibility, and persistence in the breaking condition compared to the non-breaking condition. The feeling of breaking the rules or being able to break them gives a sense of liberation, and consequently, tends to move away from the constraints of conventional mentality and explore other potential new perspectives, which could be reflected in greater cognitive flexibility (Wang et al., 2018). A piece of advice to keep in mind is the importance of maintaining a healthy lifestyle and brain health (Fourie & Fourie, 2013) and Csikszentmihalyi (1996) emphasizes the importance of this flow in the mind especially when engaging in an activity in which one is completely immersed and focused.

The concept of lateral thinking is intimately linked to creativity, as it involves reasoning that does not proceed linearly but in leaps and does not necessarily produce original ideas but, by managing to detach from the rigidity of vertical thinking, allows for the creation of alternative solutions, since it enables viewing problems from an angle other than the first that comes to mind, the one that for us is the most obvious (De Bono, 1970).

Creativity is considered a particular style of divergent thinking (Guilford, 1950) that involves generating new ideas that are of some utility or value in situations where there are many possible positive responses (Sternberg & Lubart, 1995), but it also proves to be a fundamental piece in the problem-solving process, presenting in various stages that require different skills and thinking styles (Basadur et al., 2000). The emphasis of this problem-solving process includes the final level of implementation, where the focus of interest is the moment of insight generation, typically discussed as innovation rather than the realization of creativity. Consequently, the enhancements of creativity introduced in the study by Thornhill-Miller and Dupont (2016) are those that improve some stage or aspect of the creative problem-solving process or the generation of ideas at the heart of such a process.

There is indeed no scientific way to determine which production can be considered creative and which cannot, and furthermore, there are different approaches to investigating it: some researchers have preferred to focus on the product, others on the process criterion. The process approach assumes that creativity is a trait normally distributed in the general population, while the product approach defines creativity in terms of exceptional creative output, achieved by few individuals (Brown, 1989). According to Reuter et al. (2005), however, the core of creativity, namely originality, flexibility, and elaboration, is a component that characterizes the population indiscriminately. Some scholars (Jäger, 1982; Guilford, 1967) consider creativity to be an aspect of intelligence, but, unlike the latter, there are few attempts to identify its biological correlates. Creativity can thus be seen as a specific form of problem-solving related to divergent thinking.

Heilman et al. (2003) in their review on the possible brain mechanisms of creativity, conclude that the activity of the frontal lobes is essential for creative innovation. The lobes are the primary cortical region controlling the locus coeruleus-norepinephrine (NE) system and regulate the signal-to-noise ratio, such that low levels of NE reduce the signal-to-noise ratio leading to better co-activation across modular networks, essential for divergent thinking, while high levels of norepinephrine limit the breadth of conceptual representations and increase the signal-to-noise ratio. Thus, creative individuals appear to have brains that are capable of storing extensive specialized knowledge in their temporo-parietal cortex and have a special ability to modulate the frontal-lobe locus coeruleus (norepinephrine) system such that during creativity, brain levels of norepinephrine decrease, leading to the discovery of new ordered relationships (Reuter et al., 2005).

3.4   Play and Creativity

Play activities can be used to stimulate creative activity. According to Anderson (1994), the process of play provides us with energy, concentration, and, precisely, creativity. The first key triggering element is the excitement caused by a safe risk, meaning we know we are taking a risk but that it is under control, making us feel safe as there will be no consequences. The second element is the stimulation derived from physiological activation. It is crucial that safety and stimulation are associated. Additionally, there must be an element of uncertainty in the game that increases risk, mystery, and the likelihood of developing physiological arousal, tension, and excitement: one of the main sources of stimulation in play is the creative effort to devise a strategy that addresses uncertainty. Further evidence supporting a wisely playful environment is that play implies, but also stimulates, the use of personal strategies. It is no surprise that the ideas of gamification and serious games have become increasingly popular in education and training (De Freitas & Maharg, 2011).

An important aspect of play is represented by video games, through which research has been conducted to investigate the motivations behind enjoying scary games, with particular attention to those involving virtual reality. The commercialization of VR technology has taken horror video games to high levels of immersion and presence, generating more exciting mediated experiences (Madsen, 2016). Fear can be defined as a multidimensional reaction composed of immediate emotional responses and subsequently cognitive responses to a perceived threat stimulus in the environment (Lynch & Martins, 2015). Fear itself is an instinctive reaction to threats and assists us in either confronting danger directly or avoiding it by activating appetitive and inhibitory systems (Lee & Lang, 2009). Players experience anticipatory fear while controlling their character when immersed in a highly suspenseful narrative (Lynch & Martins, 2015) and, unlike video game players, VR players face threats directly as they experience that reality, due to the powerful sense of immersion (Madsen, 2016). Consequently, the greater presence in VR allows players to increase their levels of fear. Day (2015) found that people who played a virtual reality horror game experienced more fear compared to those who played the same horror game on a traditional screen, while another study by Madsen (2016) found that players of a VR horror game exhibited higher respiratory rates, skin conductance, and heart rate variations compared to traditional video game players. However, the fear-fun relationship is inversely proportional when fear is perceived as a negative emotion (Hoffner, 2009).

Besides fear, there are two other fundamental constructs: threat resolution and excitement. Excitement plays an important role in moderating the effects of fear, and threat resolution impacts enjoyment (Zillmann, 1980), but resolution does not explain the allure of mediated fear. The self-efficacy of horror is thus defined as the belief in one's ability to endure and cope with the challenges presented in the media (Jih-Hsuan et al., 2017). In the context of video games, it is more important to believe in one's ability to successfully handle suspense and immediate threats than the actual 'resolution' of the challenges presented. If successfully enduring the horror content produces 'cognitive euphoria,' this excitement transfers the intensity of suspense into enjoyment when the game is finished. For example, when zombies surround a terrified player who has 'lost control,' the fear leads to a negative assessment of the experience. Excitement is associated with attention, consciousness, and information processing. Typically, if excitement is high, attention decreases. In the context of virtual reality horror games, a player can experience a high level of fear due to suspense while remaining calm when confronting zombies, generating low skin conductance responses (SCR), but, on the other hand, a player experiencing a low level of fear regarding the game's content might respond with higher SCRs if unexpectedly 'surprised' by zombies. Consequently, those experiencing high levels of fear and high efficacy find that high levels of excitement lead to intense fear being evaluated positively, thus generating greater enjoyment (Jih-Hsuan et al., 2017).

In virtual reality games, suspense can be generated by darkness, a lack of clues about the direction from which an attack might come, and eerie sounds. It has also been suggested that prior VR gaming experience and overall video game competence may be factors influencing enjoyment and are considered in post-experiment evaluations. The type of experience proposed for this research leverages the fear factor to stimulate physiological arousal, as a horror-survival game was designed and programmed to ensure what Anderson (1994) defined as 'safe risk': the sample thus has the opportunity to experience fear within the safety of a secure and protected context.

The earliest research on VR dates back almost forty years, but devices that use stereoscopic images, using slightly different frames for each eye like the stereoscope, began in the 1830s (Garrett et al., 2017). Exploration in this field has been accelerating in recent years with the advent of more powerful software, higher-definition graphics, and relatively inexpensive headsets like Oculus Rift or HTC Vive. The improvement of this technology increasingly allows for a realistic effect of immersion, where the user is placed in a simulated environment that seems as engaging as the real world, and the person has a specific sense of self-location, can move to explore it, interact with objects, and their actions partly determine what happens within it (Garrett et al., 2017). Indeed, virtual reality systems are characterized by more or less complex head-mounted displays and wearable touchpads, with motion tracking that allows for full immersion: essentially, if one turns, the VR will adhere to reality, so the character's head will also move in the same direction (Fig. 3.1).

3.5   Case Study: Processes of Planning in a Virtual Reality Experience—The Link Between Arousal and Problem-Solving

The research titled "Processes of Planning in a Virtual Reality Experience – The Link Between Arousal and Problem Solving" (Marchioro & Benatti, 2022) aims to study the effects of physiological activation on creativity using virtual reality. The study specifically aims to identify the most effective technique for increasing creative output (particularly focusing on the generation of many ideas, not necessarily innovative ones). This technique could potentially serve as a standalone method to unlock free production but also as an incentive for the more efficient and productive use of

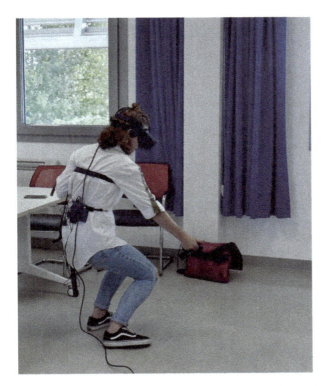

Fig. 3.1 A subject bends down to pick up an object which in the VR is placed on the ground. (Reproduced from Marchioro & Benatti, 2022)

well-established creative techniques, such as the Six Thinking Hats (De Bono, 1999), Mind Mapping (Buzan & Buzan, 2013), or SCAMPER (Eberle, 1996).

After identifying some strategies in this regard, it was hypothesized that physiological activation induced by extreme experiences could play a fundamental role in increasing creativity. Assuming this hypothesis has a basis, some critical issues quickly emerge: providing such experiences to an individual before they start working is decidedly complicated and costly, making it impractical. Therefore, the idea was to use modern technologies to recreate conditions analogous to the aforementioned experiences, leveraging the activity of mirror neurons (Rizzolatti & Sinigaglia, 2006), which can activate and evoke sensations related to actions and/or experiences that the subject does not directly experience but does so indirectly or 'vicariously.' The Oculus headset appeared to be a suitable technological tool for this purpose: it can provide firsthand experiences such as jumping into the void, climbing a rock wall, or undertaking a mission in a warzone. These experiences, in real life, usually induce a state of tension that involves the release of adrenaline (or epinephrine) by the central nervous system, a neurotransmitter that belongs to a class of substances called catechol-amines, which specifically prepare the body to face stress and danger.

The hypothesis, seemingly in contrast to the most common effects of this hormone (increased oxygen consumption, pupil dilation, increased carbon dioxide expulsion speed, reduced peripheral fatigue, increased metabolism, etc.), is that creative production could paradoxically benefit from these effects. For example, when faced with writer's block or a lack of ideas for a logo, script, or payoff, one could use an augmented reality device (the Oculus in this research) to induce physiological activation that oxygenates the mind and reactivates attention. This can be thought of as a sort of 'defibrillator' for creativity, with some important precautions: it should not be used continuously, as it would lose its functionality, and it is crucial to avoid experiencing the same virtual experience multiple times, as this would drastically reduce the necessary surprise effect (known in research as the 'trial effect').

Currently, the range of virtual experiences to apply this research is still under development: experiences characterized by emotional intensity are favored, but it is not excluded that in the future, different and initially unconsidered experiences might be identified. In summary, the project's goal is to study the effects of physiological activation on creativity through virtual reality, testing the experimental hypothesis that VR experiences with a certain emotional intensity directly contribute to stimulating cre-ativity and, consequently, problem-solving.

The reference sample comprises 24 students from the Psychology and Communication departments at IUSVE University in Venice, Italy (aver-age age: 23.36; standard deviation: 1.34; 14 females, aged 22–25 years; 10 males, aged 21–26 years), who voluntarily participated in the experi-ment. The sample was then divided into two subgroups, each consisting of 12 students, assigned to two different VR experiences: one survival-horror, designed to provoke physiological activation, and the other 'neutral,' aim-ing to limit levels of physiological arousal.

To test the subjects' creativity, the Tower of London (TOL) test, which assesses problem-solving ability, was administered, assuming a direct correlation between problem-solving and creativity as hypothesized by Weisberg (2006) and later by Golnabi (2016). Simultaneously, to measure physiological activation, the subjects were monitored using Biofeedback, which, among other indices, allows observing 'skin conductance,' directly linked to arousal. The primary hypothesis was that more stimulated subjects would demonstrate a greater ability to solve problems than those participating in the neutral experience. Contrary to expectations, the results, however, showed that physiological activation inhibits problem-solving ability, with potential effects on the participants' creative abilities.

The VR simulation was organized into four scenarios (referred to as maps in the Unreal Engine vocabulary). The operators overseeing the subjects during the experiment could select which scenario to activate, following the planned sequence.

The training scenario (Figs. 3.2 and 3.3), the first to be visualized, is the environment where the user is trained to use the virtual reality system.

The goal is to enable the user to use the touchpads for movement and interaction with accessories. The movement control through the touchpad uses a teleportation mechanism, allowing the user to decide where to

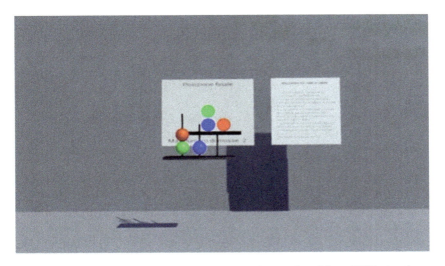

Fig. 3.2 Training scenario: the environment is populated by a TOL simulator with which to practice while the panel on the right shows the related rules. (Reproduced from Marchioro & Benatti, 2022)

Fig. 3.3 Training scenario: other objects to practice with, including a plant that will then be present in the relaxation map, a golden lion from the horror-survival map and a magazine to *insert* into the gun to try to shoot. (Reproduced from Marchioro & Benatti, 2022)

go by selecting the destination point. This solution contrasts with the classic video game mode, where movement occurs through a joystick (an input system using a lever to decide in which direction the user will move) and was chosen to avoid motion sickness, typical in video games, i.e., the sense of nausea resulting from movement in a virtual environment not matched by a corresponding physical response. Interaction with accessories is managed by training the user to grab spheres and place them on pegs in a TOL simulation, to grab ornaments and plants (as required later), and to use a gun (the latter only for users experiencing the horror scenario).

The TOL scenario (Fig. 3.4) is a scenario where the user finds themselves in a small room with movement disabled. In front of them, they can see the set of pegs and spheres that make up the tools for the test, and a panel where instructions on how to proceed and the results of the performed actions are displayed sequentially. The operator conducting the tests could initiate all items defined for that specific phase of the simulation. All test results were then saved in an Excel file used for analyses.

Fig. 3.4 TOL scenario in VR. (Reproduced from Marchioro & Benatti, 2022)

The horror scenario (Fig. 3.5) aims to scare the user to induce physiological activation. It is a survival-horror video game where the user must survive a zombie attack that appears randomly on the map. The scenario unfolds through a garden furnished with eerie statues, progressing to a dark house filled with Gothic paintings and disordered or burning objects, finally leading through a hatch to an underground tunnel that connects to an immense and deep chimney. The soundtrack consists of two dark musical pieces. The objective of the game is to collect six lion-shaped ornaments distributed throughout the scenario. If the user collects all the ornaments, they are returned to the training scenario.

The user is warned of a zombie's creation through a visual effect similar to TV frequency distortion followed by a series of loud sounds. The zombie's placement can be near the user or elsewhere. The attack occurs through hand strikes: if five strikes are successful, the user dies and is brought back to the training scenario. If the zombie is placed near the user, it attacks immediately, producing a loud, distinct sound. If placed elsewhere, it waits until the user enters its field of view, then proceeds to attack. The user can defend themselves by killing the zombies using a gun (Fig. 3.6). The entire gameplay phase should not last more than five minutes, with the operator keeping track of time and interrupting it if necessary, bringing the user back to the training scenario.

In contrast, the relaxation scenario aims to provide the user with a neutral situation that does not provoke significant physical effects. It is a very

Fig. 3.5 Scenario of the survival-horror game, in the foreground you can see the gun with which the zombies are fought. (Reproduced from Marchioro & Benatti, 2022)

Fig. 3.6 Zombie attacking: in the foreground you can see the gun that fired shots to destroy it. (Reproduced from Marchioro & Benatti, 2022)

simple video game set in a meadow surrounded by high rocks, featuring trees, plants, and flowers (Fig. 3.7).

The soundtrack is relaxing music. The objective of the game is to collect six golden plants scattered across the meadow. Once all the plants are collected, the user wins and is returned to the training scenario. There is no option for the user to die. As with the horror scenario, the entire gameplay phase should not last more than five minutes, with the operator keeping track of time and interrupting it if necessary, bringing the user back to the training scenario.

The tests were conducted over two days, simultaneously using two stations in adjacent rooms. The first Oculus device was used to administer the test to the experimental group, while the second was used for the control group. Each subject required approximately 35–40 minutes, depending on the time taken to complete the Tower of London (TOL) test.

Students who participated in the project as experimental subjects were seated in a nearby room to relax and concentrate and were called individually in order of arrival. Once the subject entered the experimental room, they were alone with the biofeedback operator and the virtual reality device operator to avoid discomfort and eliminate any potential

Fig. 3.7 Neutral and natural environment. (Reproduced from Marchioro & Benatti, 2022)

distractions. This setup ensured proper supervision of the experiment and monitoring of the subject's movements to maintain safety.

First, the biofeedback electrodes were applied to the subject, their function was explained, and psychophysiological parameters were measured. Subsequently, the biofeedback cables were disconnected while keeping the sensors in place. The subject was then fitted with the VR headset and touchpads and underwent training to familiarize themselves with the Oculus functions and commands. Once the subject understood the instructions, the TOL rules were explained through the virtual display of an information panel.

The first part of the TOL test was then conducted to measure problem-solving abilities used to address the proposed problems. There were twelve items in total, divided as follows: odd-numbered items in the first part and even-numbered items in the second. After completing the first six items, subjects in the experimental group underwent the experience designed to trigger physiological activation, while those in the control group had a neutral experience. Both experiences utilized virtual reality. Five minutes were allocated for both experiences unless the tasks were completed before the time limit. Following this part, the subject proceeded with the even-numbered items of the TOL test and concluded with a paper questionnaire.

Regarding skin conductance, there was a noticeable decrease in physiological activation in both groups, though with some significant differences: specifically, the control group's average decreased from 1.302 (pre-treatment) to 1.186 (post-treatment), while the experimental group's average decreased from 2.416 (pre-treatment) to 2.202 (post-treatment) (Table 3.1).

The independent samples T-test shows that the conductance values do not differ significantly between the two groups, neither before nor after the treatment, although they are higher in the experimental group in both cases (Pre: $t = -1.939$, Cohen's $d = -0.792$; Post: $t = -1.916$, Cohen's $d = -0.782$).

Below, we report the mean values, standard deviation, minimum, and maximum values related to the main dependent variable considered, namely the scores obtained by the subjects in the 6 TOL items, both before and after the treatment (Table 3.2).

We note that in both cases, there is a drastic drop in performance, linked to the decrease in scores obtained by the subjects after the treatment. Conducting a paired samples T test show that this drop in

Table 3.1   Skin conductance after and before TOL (Reproduced from Marchioro & Benatti, 2022)

	Skin_Cond_uS_Pre_TOL		Skin_Cond_uS_Post_TOL	
	Control	Experimental	Control	Experimental
Valid	12	12	12	12
Missing	0	0	0	0
Mean	1.302	2.416	1.186	2.202
Std. Deviation	1.236	1.558	1.214	1.379
Minimum	0.2000	1.110	0.1700	1.080
Maximum	4.310	6.010	4.370	4.960

Table 3.2   Mean, standard deviation, minimum and maximum relating to the scores obtained by the subjects in the 6 items of the TOL, both before and after the treatment (Reproduced from Marchioro & Benatti, 2022)

	TOL_Pre		TOL_Post	
	Control	Experimental	Control	Experimental
Valid	12	12	12	12
Missing	0	0	0	0
Mean	3.833	3.583	1.833	1.833
Std. Deviation	2.125	1.730	1.697	1.586
Minimum	0.000	0.000	0.000	0.000
Maximum	6.000	6.000	6.000	6.000

performance, regardless of distinguishing between the experimental and control groups, is even significant ($t = 4.577$, with $p < 0.001$ and Cohen's $d = 0.934$), as displayed in the table below. This suggests that, regardless of the type of experience, all subjects seem to have suffered a performance decline.

It was considered that what was observed might depend on potential performance differences between males and females. Therefore, an independent samples T-test was conducted to dispel this doubt: however, the results, as reported in the table below, showed the absence of any differences between the two groups (respectively males and females), both before and after the treatment (pre-$t = -0.231$, with $p = 0.819$ and post-$t = 0.854$, with $p = 0.402$).

To verify the effects of the treatment, paired samples T-tests were conducted to search for any differences within each group (experimental vs. control) between the measurements obtained before and after the treatment. There are no significant differences between the two groups in any of the considered variables. In particular, this means that, on average, execution and reaction times, following the experience, are similar or, in any case, not so different as to justify a potential treatment effect.

In the experimental group, while the level of activation remains almost unchanged, there is a significant drop in performance, especially related to the TL variable: in this case, the group goes from an average of 3.583 to an average of 1.833 ($t = 3.023$, with $p < 0.05$). Similarly, there is an increase in the number of moves and, consequently, attempts between before and after. To verify potential differences in reaction and execution times before and after the treatment within the Experimental Group, a paired samples T-test (repeated measures) was conducted. There is a significant increase in reaction time (latency between stimulus and decision) in the subjects of the Experimental Group: indeed, it goes from an average of 7.56 seconds to an average of 11.33 seconds ($t = -3.26$, with $p < 0.01$). Regarding execution times, however, there are no significant differences between the two recorded values: this means that execution speed, contrary to reaction times to stimuli, is not influenced by the VR experience.

In the control group, it emerges that the drop only concerns the score obtained in the TOL test, as it goes from an average of 3.833 before the treatment to an average of 1.833 after the treatment ($t = 3.317$, with $p < 0.01$). The drop in the control group is therefore more significant, in this variable, than observed in the experimental group. The same analysis procedure was adopted to verify the significance of changes in reaction and execution times within the Control Group. In the experimental group, there are no significant differences between reaction times or execution times.

The study, born from a hypothesis not yet sufficiently explored in existing literature (through the use of the Oculus Rift, biofeedback procedure, and the VR version of the Tower of London test), analyzed the possible correlation between physiological activation and problem-solving. Two virtual reality experiences were developed to relate the results of the aforementioned psychological test of the experimental group with those of the control group. Following the experiments conducted over two days, the results were collected, cleaned, and analyzed.

It emerged that, contrary to expectations, the entire sample experienced a drop in the Tower of London test results, especially the experimental group, both in terms of reaction times and the significant increase in errors and moves used to complete the tests. This could confirm the theory that high levels of norepinephrine limit the breadth of conceptual representations by increasing the signal-to-noise ratio, resulting in poorer co-activation across modular networks essential for divergent thinking (Heilman et al., 2003). There is a possibility that physiological activation, in situations requiring reasoning rather than impulsive action (where adrenaline comes into play), worsens performance. It slows down and distracts the subjects. This could be a starting point from which to develop new pathways to explore this direction further.

An element that distinguishes the research and denotes novelty in the psychological field is the conversion of the Tower of London test into virtual reality, approved by professionals in the field before use, confirming its realism. The choice was made to simplify the experiment without repeatedly removing and reattaching the headset and touchpads, thereby promoting subject concentration. This first 'trial' opens a gap between two very different worlds, introducing potential new hypotheses regarding the effects of virtual reality on human psychology.

The inability to use wireless tools, both for the Oculus and biofeedback, proved to be a significant obstacle. At the time of the research, new wireless versions were not yet available, preventing the biofeedback system from recording data during the test due to the necessary but incompatible movements with the cables. This compromised part of the result as subject parameters were only recorded before and after the test, consequently losing fundamental data for interpreting the outcome or highlighting further changes. Surely, in the perspective of a future upgrade of the experiment, it will be necessary to use advanced systems capable of supporting situations where the subject is not sitting still but moves in the six directions mentioned in the second chapter without hindering psychophysiological parameter recording.

Similarly, in anticipation of further experiments, it will be possible to consider using the wireless Oculus Quest, which will allow a preliminary study (conducted by the device itself) of the surrounding spaces, eliminating the need for an operator to protect the subject from potential collisions with nearby objects and allowing unrestricted movement (except for real barriers) within the space. However, an important constraint to consider regarding all-in-one VR devices is that the graphic quality cannot yet compare with that of devices based on an external computer, such as the Oculus Rift. Sometimes this is not a problem as it does not need to appear 'real,' but in the case of experiences requiring high quality for greater impact, it could be a research limitation.

Another limitation was the voluntary participation of the sample. As explained in the second chapter, subjects were recruited in person or through an announcement: the fact that they deliberately decided to undergo the test may have influenced the experiment's outcome, as there is a possibility they had an interest in participating, making willingness a variable to consider in the result analysis. To limit result variability, it is deemed essential that the sample be recruited to be correctly representative of the observed population, and consequently, subject selection should be done according to a statistical proportion.

An additional limitation concerns the number of subjects: it was initially planned for the control group to be divided into two subgroups experiencing different tests. For the group that could not be involved in this research for logistical reasons, a non-VR experience was planned, proposing the viewing of a National Geographic video using the YouTube platform to exclude potential 'falsifications' of the results by considering VR itself as a variable, regardless of the experience. In the future, the sample must be significantly larger to allow comparisons between subgroups in analyzing the results derived from the various employed tools.

References

Anderson, J. V. (1994). Creativity and play: A systematic approach to managing innovation. *Business Horizons, 37*(2), 80–86. Amsterdam: Elsevier.

Basadur, M., Runco, M. A., & Vega, L. A. (2000). Understanding how creative thinking skills, attitudes and behaviors work together: A causal process model. *Journal of Creative Behavior, 34*(2), 77–100. New Jersey: Wiley-Blackwell.

Botella, C., Serrano, B., Baños, R. M., & Garcia-Palacios, A. (2015). Virtual reality exposure-based therapy for the treatment of post-traumatic stress disorder: A review of its efficacy, the adequacy of the treatment protocol, and its acceptability. *Neuropsychiatric Disease and Treatment, 11*, 2533–2545. https://doi.org/10.2147/NDT.S89542

Bowers, K. S., Regher, G., Balthazard, C., & Parker, K. (1990). Intuition in the context of discovery. *Cognitive Psychology, 22*, 72–110.

Bransford, J. D., & Stein, B. S. (1993). *The ideal problem solver: A guide for improving thinking, learning and creativity.* W.H. Freeman.

Brown, R. T. (1989). Creativity: What are we to measure? In J. A. Glover, R. R. Ronning, & C. R. Reynolds (Eds.), *Handbook of creativity* (pp. 3–32). Plenum Press.

Buzan, T., & Buzan, B. (2013). *Mappe mentali: Come utilizzare il più potente strumento di accesso alle straordinarie capacità del cervello per pensare, creare, studiare, organizzare.* Unicomunicazione.it.

Cavallin, F. (2015). *Creatività, pensiero creativo e metodo.* libreriauniversitaria. it edizioni.

Csikszentmihalyi, M. (1996). *Creativity: Flow and the psychology of discovery and invention.* Harper Perennial.

Davidson, J. E., & Sternberg, R. J. (1984). The role of insight in intellectual giftedness. *Gifted Child Quarterly, 28*(2), 58–64.

Day, T. W. (2015). *The Oculus rift as a portal for presence: The effects of technology advancement and sex differences in the horror video game genre.* Doctoral Dissertation, Michigan State University.

De Bono, E. (1970). *Lateral thinking: A textbook of creativity.* Mica Management Resources.

De Bono, E. (1999). *Sei cappelli per pensare.* Rizzoli.

De Freitas, S., & Maharg, P. (2011). *Digital games and learning.* Continuum International.

Dul, J., & Ceylan, C. (2011). Work environments for employee creativity. *Ergonomics, 54*(1), 12–20. https://doi.org/10.1080/00140139.2010.542833

Duncker, K. (1945). On problem-solving. *Psychological Monographs, 58*(5).

Eberle, B. (1996). *Scamper: giochi per lo sviluppo dell'immaginazione.* Prufrock Press.

Fourie, I., & Fourie, H. (2013). Getting it done on time. *Library Hi Tech, 31*(2), 391–400. Bingley: Emerald Group Publishing Limited.

Garrett, B., Taverner, T., & McDade, P. (2017). Virtual reality as an adjunct home therapy in chronic pain management: An exploratory study. *JMIR Medical Informatics, 5*(2), e11. https://doi.org/10.2196/medinform.7271

Golnabi, L. (2016). Creativity and insight in problem solving. *Journal of Mathematics Education at Teachers College, 7*(2), 27–29.

Greeno, J. G. (1978). Natures of problem-solving abilities. In W. K. Ertes (Ed.), *Handbook of learning and cognitive processes* (Vol. 5). Erlbaum.

Guilford, J. P. (1950). Creativity. *American Psychologist, 5*(9). Washington: American Psychological Association.

Guilford, J. P. (1967). *The nature of human intelligence.* McGraw-Hill.

Hayes, J. R. (1989). *The complete problem solver.* Erlbaum.

Heilman, K. M., Nadeau, S. E., & Beversdorf, D. O. (2003). Creative innovation: Possible brain mechanism. *Neurocase, 9*(5), 369–379. Regno Unito: Taylor & Francis.

Hoffner, C. (2009). Affective responses and exposure to frightening films: The role of empathy and different types of content. *Communication Research Reports, 26*(4), 285–296. Abingdon-on-Thames: Routledge.

Jäger, A. O. (1982). *Berliner Intelligenzstruktur-Test (BIS-Test)*. Hogrefe.

Jih-Hsuan, T. L., Dai-Yun, W., & Chen-Chao, T. (2017). So scary, yet so fun: The role of self-efficacy in enjoyment of a virtual reality horror game. *New Media & Society, 20*(1), 3223–3242. California: SAGE Publications.

Köhler, W. (1917). *Intelligenzprüfungen an Anthropoiden*. Abhandlungen d. Kgl. Preuss. Akad. d. Wiss. Phys.-math. Kl.

Lee, S., & Lang, A. (2009). Discrete emotion and motivation: Relative activation in the appetitive and aversive motivational systems as a function of anger, sadness, fear, and joy during televised information campaigns. *Media Psychology, 12*(2), 148–170. Boston: Hogrefe Publishing Group.

Leung, A. K., et al. (2012). Embodied metaphors and creative "acts". *Psychological Science, 23*(5), 502–509. California: SAGE Publications.

Lynch, T., & Martins, N. (2015). Nothing to fear? An analysis of college students' fear experiences with videogames. *Journal of Broadcasting & Electronic Media, 59*(2), 299. Abingdon-on-Thames: Routledge.

Madsen, K. E. (2016). The differential effects of agency on fear induction using a horror-themed video game. *Computers in Human Behavior, 56*(3), 142–146. Amsterdam: Elsevier.

Magro, T., & Muffolini, E. (2011). *Fondamenti di psicologia generale* (Vol. 1). LED.

Marchioro, G., & Benatti, F. (2022). *Processes of planning in a virtual reality experience: Link between arousal and problem solving*. Poster presented at the *17th European Congress of Psychology (ECP) "Psychology as the Hub Science: Opportunities and Responsibility"—Lubiana (Slovenia)*.

Metcalfe, J., & Wiebe, D. (1987). Intuition in insight and non-insight problem solving. *Memory and Cognition, 15*, 238–246.

Nechvatal, J. (2009). *Immersive ideals/critical distances*. Lambert Academic Publishing.

Oppezzo, M., & Schwartz, D. L. (2014). Give your ideas some legs: The positive effect of walking on creative thinking. *Journal of Experimental Psychology: Learning Memory and Cognition, 40*(4), 1142. Washington: American Psychological Association.

Reed, S. K. (1988). *Cognition: Theory and application*. Wadsworth. (Trad. it.: *Psicologia cognitiva*. Bologna: Il Mulino, 1989).

Reuter, M., et al. (2005). Personality and biological markers of creativity. *European Journal of Personality, 19*(2), 83–95. New Jersey: John Wiley & Sons.

Rizzolatti, G., & Sinigaglia, C. (2006). *So quel che fai: Il cervello che agisce e i neuroni specchio*. Raffaello Cortina Editore.

Slater, M. (1999). Measuring presence: A response to the Witmer and Singer presence questionnaire. *Presence, 8*(5), 560–565.

Slater, M. (2003). A note on presence terminology. *Presence Connect, 3*(3), 1–5.

Slater, M., Usoh, M., Linakis, V., & Kooper, R. (1996, July). Immersion, presence and performance in virtual environments: An experiment with tri-dimensional chess. In *Proceedings of the ACM symposium on virtual reality software and technology* (pp. 163–172).

Slepian, M. L., & Ambady, N. (2012). Fluid movement and creativity. *Journal of Experimental Psychology: General, 141*(4), 625. Washington, American Psychological Association.

Slepian, M. L., Weisbuch, M., Rutchick, A. M., Newman, L. S., & Ambady, N. (2010). Shedding light on insight: Priming bright ideas. *Journal of Experimental Social Psychology, 1*(46(4)), 696–700. https://doi. org/10.1016/j.jesp.2010.03.009

Sternberg, R. J., & Lubart, T. I. (1995). *Defying the crowd: Cultivating creativity in a culture of conformity* (Vol. 9). Free Press.

Szabó, B. K., & Gilányi, A. (2020). The notion of immersion in virtual reality literature and related sources. In *2020 11th IEEE international conference on cognitive infocommunications (CogInfoCom)*, Mariehamn (pp. 000371–000378). https://doi.org/10.1109/CogInfoCom50765.2020.9237875.

Thornhill-Miller, B., & Dupont, J.-M. (2016). Virtual reality and the enhancement of creativity and innovation: Under recognized potential among converging technologies? *Journal of Cognitive Education and Psychology, 15*(1), 102–121. New York, Springer Publishing.

Turner, W. A., & Casey, L. M. (2014). Outcomes associated with virtual reality in psychological interventions: Where are we now? *Clinical Psychology Review, 34*(8), 634–644. https://doi.org/10.1016/j.cpr.2014.10.003

Wang, X., Lu, K., Runco, M. A., & Hao, N. (2018). Break the "wall" and become creative: Enacting embodied metaphors in virtual reality. *Consciousness and Cognition, 62*, 102–109. Amsterdam, Elsevier.

Weisberg, R. W. (2006). *Creativity: Understanding innovation in problem solving, science, invention, and the arts.* John Wiley & Sons.

Wertheimer, M. (1945–1959). *Productive thinking.* Harper.

Witmer, B. G., & Singer, M. J. (1998). Measuring presence in virtual environments: A presence questionnaire. *Presence, 7*(3), 225–240.

Zillmann, D. (1980). Anatomy of suspense. In P. H. Tannenbaum (Ed.), *The entertainment functions of television.* Lawrence Erlbaum Associates.

CHAPTER 4

Cognitive Fatigue and Restorative Effects of Virtual Natural Environments

Abstract This chapter delves into the phenomenon of cognitive fatigue, examining its definitions, classifications, and significant impacts on mental performance and psychological well-being. Acute Cognitive Fatigue (ACF), a severe mental exhaustion resulting from demanding cognitive tasks, is explored alongside measurement methods such as n-back tasks and Raven's matrices. A key focus is on the restorative effects of natural environments, guided by the Attention Restoration Theory (ART). The theory posits that interactions with nature can facilitate cognitive recovery, enhance mental focus, and reduce stress. This section also addresses the detrimental effects of urbanization and environmental degradation on mental health, emphasizing the therapeutic benefits of natural settings. In light of limited access to natural environments, the chapter introduces Virtual Reality (VR) as an innovative tool for simulating restorative natural experiences. Empirical evidence is presented demonstrating VR's effectiveness in promoting psychological well-being and cognitive restoration, especially for individuals unable to engage with real natural settings.

A case study is included, illustrating the application of VR in creating restorative environments. The study compares the effects of high and low restorative potential images in 3D and 2D formats on cognitive fatigue, highlighting the significant restorative potential of immersive VR environments. The results indicate that high-quality, immersive VR environments substantially reduce cognitive fatigue, with 3D images showing greater restorative effects than 2D images.

Keywords Virtual reality • Acute cognitive fatigue (ACF) • Attention restoration theory (ART) • VR environments • Restorative environments

4.1   UNDERSTANDING COGNITIVE FATIGUE

When considering the concept of *fatigue*, the first thought often connects to "tiredness", understood as a condition of physical and/or mental exhaustion characterized by a decrease in energy, manifesting episodically or persistently. Literature defines fatigue as a set of cognitive, emotional, and physical symptoms that are unpleasant and not alleviated by usual strategies for recovering psychophysiological resources (Mota & Pimenta, 2006).

In our active and incessant society, which operates 24/7, the issue of fatigue assumes increasing urgency (Harrington, 2012; Argüero-Fonseca et al., 2023a). Early studies on the subject revealed that between 5% and 9% of the global population experiences episodes of fatigue lasting at least six months (Pawlikowska et al., 1994), with variations in incidence across different work sectors and a higher prevalence among women compared to men, at a ratio of 4:1 (Afari & Buchwald, 2003).

Fatigue can be classified in terms of duration: recent (less than one month), prolonged (from one to six months), and chronic (over six months) (Cornuz et al., 2006). Depending on the duration and intensity, this condition can have significant impacts on quality of life and work productivity, leading to functional limitations, disability, and in severe cases, an increased risk of mortality (Argüero-Fonseca et al., 2023a).

Cognitive fatigue is a concept widely studied in relation to mental performance and psychological well-being. Fatigue is conceived as a multidimensional construct, encompassing a wide range of both physical and mental determinants. Despite numerous studies conducted on specific populations (Bültmann et al., 2002; David et al., 1990; Fuhrer & Wessely, 1995), research concerning the general population has yet to exhaustively explore the causes of this condition, leaving many aspects still to be investigated (Galland-Decker et al., 2019).

The term "cognitive fatigue" thus refers to a specific manifestation of fatigue, defined as acute mental fatigue. This phenomenon, linked to mental performance, manifests following the prolonged execution of tasks that require intense cognitive processing. The concept of acute mental fatigue is frequently utilized in literature as synonymous with mental fatigue and/or acute cognitive fatigue (Bafna et al., 2021; Argüero-Fonseca et al., 2023a).

4.1.1 Dynamics and Implications of Acute Cognitive Fatigue (ACF)

As previously emphasized, cognitive fatigue is a complex phenomenon that involves the wear of the attentive state, linked to both subjective conditions, identified as subjective fatigue, and objective conditions, known as *fatigability*. These conditions emerge as a result of a gradual and cumulative modification of the psychophysiological state, typically caused by the prolonged execution of tasks that require constant and sustained attention.

The definition of cognitive fatigue embraces the complexity of acute mental fatigue and extends to include the behavioral, perceptual, and physiological manifestations that occur during and after cognitively demanding tasks. This specific form of mental fatigue, distinct from chronic forms of fatigue or cognitive impairment related to age or diseases, includes a range of subjective sensations such as profound tiredness, lack of energy, desire for rest, and a decrease in motivation, which can manifest during or at the end of prolonged periods of either monotonous or highly demanding cognitive activities (Hopstaken et al., 2015).

Acute cognitive fatigue is characterized by a sensation of extreme tiredness, often accompanied by an aversion to continuing the current task and a reduction in effort. This state can culminate in a significant decrease in the level of sustained attention and working memory capacity, highlighting how fatigue affects not only psychological well-being but also the cognitive and physical performance of the individual (Boksem & Tops, 2008; Faber et al., 2012). These indicators of fatigability reflect the physical and behavioral manifestations associated with cognitive fatigue and underline the importance of considering these aspects in assessing and managing acute mental fatigue.

4.1.2 Assessing Cognitive Fatigue: Methods and Tools

Measuring cognitive fatigue is an area of growing interest in psychology and neuroscience, as cognitive capacity can be significantly affected by conditions of mental fatigue. Cognitive fatigue can thus be induced through cognitive tasks that require high levels of concentration and prolonged mental effort. Tools commonly used to assess this phenomenon include the n-back task and Raven's matrices.

The *n-back* task, primarily used as a tool for assessing working memory (Jaeggi et al., 2010), proves particularly effective in inducing cognitive fatigue by requiring participants to monitor a sequence of stimuli and respond when a stimulus matches the one presented n positions back, thus increasing the workload on working memory. The results of a recent study, in which cognitive fatigue was induced and analyzed through the use of the n-back task, show that cognitive fatigue can be measured through various physiological and behavioral indicators, such as increased response latency and decreased accuracy in performance (Argüero-Fonseca et al., 2023b). These effects have also been highlighted in previous studies, which have shown that excessive demands on working memory can lead to a significant reduction in attention capacity and overall performance (Guifang et al., 2017; Smith et al., 2019).

Raven's matrices, on the other hand, measure fluid intelligence and problem-solving ability through the resolution of visuo-spatial problems, requiring complex and sustained reasoning processes (Raven, 2000). The use of these tests allows the quantification of cognitive fatigue by observing the decline in performance over time and correlating these decreases with subjective measurements of tiredness reported by participants (Lorist et al., 2005). Empirical evidence thus suggests that monitoring variations in cognitive performance can provide a robust measure of cognitive fatigue, contributing to the understanding of its effects on overall cognitive functions.

4.2   ENVIRONMENTAL IMPACT ON MENTAL FATIGUE RECOVERY

In the contemporary context, there is a growing awareness of the significant impact that the environment has on the psychological well-being of individuals. This understanding has catalyzed the development of new interdisciplinary branches of study, including *Environmental Psychology*. This discipline was conceptually outlined by Stokols (1990) as a field of inquiry focused on the bidirectional interactions between humans and their environmental contexts. Stokols particularly emphasized how the perceptions, cognitive processes, and actions of individuals not only are influenced by the environment but also contribute to its modification (Stokols, 1990).

Studies by Kaplan (1995) have demonstrated the regenerative capacity of nature, revealing how even brief periods of interaction with natural environments can induce significant therapeutic benefits, including stress reduction and the enhancement of psychological resilience (Kaplan, 1995). These positive effects also extend to other dimensions, such as increased emotional serenity and heightened creative productivity (Kaplan, 1995).

In an era marked by relentless urbanization and a progressive distancing of human settlements from the natural environment, there is an increasingly frequent occurrence of harmful effects on the mental health of individuals (Kellert & Wilson, 1993). This phenomenon, known as *environmental alienation*, has over time become the subject of scientific scrutiny, with several authors highlighting how detachment from the natural environment translates into a lower quality of life, with severe repercussions on physical and mental well-being: the decline in biodiversity, climate changes, and environmental deterioration are indeed factors that influence not only ecosystems but also the psychological well-being of people, thereby disturbing mental balance and individual serenity (Doherty & Clayton, 2011). Natural disasters, increasingly frequent and devastating due to climate changes, such as hurricanes, fires, and floods, have been shown to have profound and lasting impacts on the mental health of affected communities. Fritze et al. (2008) explored the psychological consequences of such events, highlighting how the repercussions extend beyond immediate physical damage, lasting over the long term and requiring targeted psychosocial interventions (Fritze et al., 2008).

Emerging from this research context, the contribution of Gifford (2014) represents a significant milestone, drawing on previously emphasized points, and significantly bridging psychology with the surrounding natural environment. He explores the importance of environmental psychology as a discipline that studies the interactions between individuals and their environments, both natural and constructed, analyzing how these environments influence human behavior and vice versa, to highlight behaviors that promote or hinder environmental sustainability. Gifford's work emphasizes the importance of interventions designed to increase pro-environmental behaviors, underlining the importance of environmental policies and sustainable design practices to improve human and environmental well-being (Gifford, 2014). Furthermore, Gifford's analysis underlines how ecosystem destruction and degradation can have direct repercussions on people's mental health, emphasizing the need for political and practical action to reform the relationship between humans and the environment.

Research in this field has explored various areas, including the influence of urban green spaces on recovery from stress and the enhancement of mental health (Hartig et al., 2014). The interdisciplinarity between psychology and ecology has further enriched this field of study, expanding research methodologies and practical applications in areas such as urban planning, environmental conservation, and the support of psychological well-being.

In recent years, interest in the dynamic interaction between individuals and the surrounding environment has grown considerably. This interest is fueled by the increasing concern among mental health experts regarding the psychological implications of accelerated urbanization and the global ecological crisis. These concerns, as already highlighted, become even more relevant in an era characterized by a hectic lifestyle and pervasive urbanization (Marchioro, 2023).

In the outlined context, the importance of the already discussed concept of cognitive fatigue becomes clearer. With the increase in urban stimuli and the reduction of contact with nature, individuals may experience an information overload that leads to a long-term decrease in cognitive capacities. This state predominantly manifests through difficulties in concentration, mental exhaustion, and reduced capacity to process new information. Recognizing and understanding cognitive fatigue thus becomes crucial for developing effective mitigation strategies that include cognitive restoration through exposure to natural environments, aimed at promoting concrete actions to reduce the negative impacts of urbanization (Marchioro, 2023). In this way, not only individual well-being could be supported but also collective well-being, promoting environments that foster psychological and cognitive resilience in response to modern environmental stresses.

4.2.1   Attention Restoration Theory (ART)

In the late 1980s and early 1990s, a period marked by rapid technological progress and an increase in indoor activities, such as prolonged office work and the spread of indoor entertainment, Stephen Kaplan and Rachel Kaplan developed the *Attention Restoration Theory (ART)*. This theory provides a framework for analyzing the relationship between recovery from mental fatigue and interaction with the natural environment. The Kaplans investigated the impact of natural environments on the human psyche, focusing particularly on what makes such environments particularly compelling (Kaplan & Kaplan, 1989).

The results of their research led to the formulation of ART, which suggests that exposure to nature is not only inherently pleasing but also plays a crucial role in enhancing attention and concentration abilities (Ohly et al., 2016). ART is based on the premise that human cognitive capacities have evolved within natural environments and is distinguished by its emphasis on the duality between directed attention, which requires conscious effort, and fascination, a state of immersive and spontaneous engagement (James, 1892). Thus, the theory not only deepens our understanding of the interaction between humans and nature but also highlights the vital importance of natural environments in supporting cognitive functions.

4.2.1.1 The Restorative Function of Natural Environments: A Four-Phase Process

Academic literature highlights numerous benefits derived from exposure to natural environments (Hartig et al., 2014; Argüero-Fonseca et al., 2023b; Kaplan, 1995). Kaplan and Kaplan (1989) argue that these benefits are intrinsically connected to the aesthetic attributes offered by nature. Staying in natural spaces is not only pleasant but also facilitates effective management of information from the external environment, allowing for safe and comfortable exploration. However, the primary psychological advantage lies in the ability of natural environments to accelerate recovery from mental fatigue, promoting the renewal of cognitive energy (Kaplan & Kaplan, 1989).

The restorative function of natural environments is particularly complex and unfolds in four distinct phases:

1. *Clearing the Head*: after intense and demanding cognitive activities, the mind can remain cluttered with cognitive residues such as thoughts, words, and images that hinder the initiation of new activities. Natural environments, free from excessive stimuli, allow these residues to naturally dissolve, freeing the mind from cognitive obstructions.

2. *Mental Fatigue Recovery*: during this phase, there is a recovery of mental energy as attention shifts away from focused tasks toward the environment, offering a break from the usual intense stimuli.

3. *Soft Fascination*: the presence of non-invasive environmental stimuli allows for internal reflection. These elements of "soft fascination" help divert the mind from the internal noise that can emerge during cognitive activities, thus promoting personal reflection and introspection.

4. *Reflection and Restoration*: Kaplan and Kaplan observed that a truly restorative experience can include deep reflections on personal life, actions, goals, and priorities. Spending prolonged periods in a relaxing environment allows an individual to reflect on life goals and reorient efforts toward those that truly contribute to achieving their aspirations (Kaplan & Kaplan, 1989).

An experience is qualified as deeply restorative only if an individual, by moving through these four phases, manages to clear the mind of cognitive residues and fully dedicate themselves to personal reflection and growth.

4.2.1.2 Key Determinants of Restorative Environments

Kaplan and Kaplan have highlighted that daily functioning in modern environments often requires individuals to focus their attention on specific and well-defined stimuli (Kaplan & Kaplan, 1989). However, constant use of this directed attention capacity can lead to its weakening over time. This weakening can result in a series of negative effects, such as the onset of negative emotions, irritability, a reduced ability to perceive signals in interactions with others, a decrease in willingness to help others, poor performance in tasks requiring high concentration, and an increase in accidents.

To mitigate these effects, it is important to recharge our directed attention capacity. This can be done by immersing oneself in environments that do not require constant concentration effort but rather activate a more relaxed mode of attention, known as "fascination." This type of attention is less demanding and helps relax the mind. Fascination, however, is just one of four essential elements that promote recovery and restoration through experiences, whether they occur in nature or other environments. Below are all the key factors:

Being Away. Corresponds to the sensation of being distant from everyday thoughts and worries (Herzog et al., 2003). Kaplan and Kaplan (1989) describe three ways to experience this sense of detachment:

escaping from unwanted distractions in the surrounding environment, distancing oneself from usual work and its memories, and suspending the pursuit of specific goals. However, achieving such a state does not necessarily require physical removal from one's usual environment. In fact, Scopelliti and Giuliani (2004) observed that often a mental change is more restorative than a physical change (Scopelliti & Giuliani, 2004). In everyday life, it is indeed possible to experience the sense of "being away" without changing location, for example when immersed in watching a movie, reading a book, or listening to a podcast; in moments like these, it is easy to feel part of another environment, distant from one's life or worries. Being away is defined by Ragsdale et al. (2011) as "psychological detachment," within the Psychological Recovery Theory (PRT), namely the psychological experience of mental disengagement from demands during leisure time, which involves distraction from task-related thoughts (Ragsdale et al., 2011). Being away from the demands and commitments of everyday life, along with engaging in activities different from daily ones, allows the recovery from the depletion of cognitive resources and the daily accumulated stress. It is thus evident that the individual needs to be distant from the environment causing the depletion of attentive resources in order to recover them, but as observed, it is not necessary that the distance be physical; rather, it is sufficient that it be mental for this to occur.

Fascination. An activity is defined as fascinating when it can be followed without demanding attention and without suppressing other types of competing stimuli. This activity allows for the recovery of directed attention as well. According to Kaplan (1995), fascination can be divided into two types:

- *Hard fascination*: engaging fascination typical of activities with high stimulation that capture attention without leaving room for reflection.
- *Soft fascination*: gentle fascination typical of low-stimulation activities, such as observing nature, which aids in restoring attention and allows for reflection.

In both cases, attention is captured automatically, without effort from the individual, but the restorative effects can differ. "Hard fascination" experiences do not allow time for reflection and are thus useful for reducing boredom and entertaining the individual, while "soft fascination" experiences allow the individual to think more freely, providing a means to make sense of their past experiences and reflect on recent ones. Studies have shown that natural environments like gardens, forests, and beaches are more likely to elicit soft fascination and have a high potential for reflection and recovery (Herzog et al., 2003).

Extent. According to Kaplan (2001), extent refers to the type of environments that have a scope and coherence that allow an individual to feel immersed and engaged within them. Within such an environment, events unfold in a relatively simple, predictable, and orderly manner; however, this presupposes that the individual must be familiar with the environment and not be disoriented in new or unusual scenarios (Kaplan, 2001). For example, while walking a favorite mountain trail, one knows the duration of the walk, the direction to take, the level of difficulty, and the terrain's layout, all important characteristics that can be encapsulated in a sort of cognitive map. According to Kaplan (2001), cognitive maps are mental structures generated by the brain from concepts and experiences encountered throughout one's life, comprising the set of information one possesses when undertaking a particular action or moving to a particular place. With this information, it is possible to predict and anticipate what might happen, informed by past experience, and prepare for new situations. It should be noted that there are no environments and situations with isolated characteristics; rather, they often have overlapping characteristics. Consider, for instance, driving a car to work: in this case, multiple cognitive functions are activated simultaneously (driving the car, directing attention to the road, other cars, and road signs), which involves mental fatigue. It is also possible that an individual may tire when trying to inhibit other mental maps, perhaps not relevant to the current situation. Extent is thus evoked by all those environments and situations in which the individual is not forced to use a large number of cognitive maps, such as familiar environments. In these environments, the restoration of mental fatigue is more likely because there is no need to inhibit other cognitive maps, and attention is activated effortlessly. If unfamiliar stimuli or experiences arise, other cognitive maps are activated, nullifying the restorative properties of the environment.

Compatibility. This factor refers to the correspondence between an individual's goals and inclinations, the demands made by environmental conditions, and the information patterns available to support these targeted demands (Kaplan & Talbot, 1983). It is that feeling of pleasure congruent with the type of environment that, to be rejuvenating, must be exactly where the individual chooses to be, for an intrinsic motivation or personal preference; an individual present in an environment for extrinsic reasons is likely not to experience any type of restoration. Compatibility occurs when what the individual wants to do matches what the environment offers and supports. It is hypothesized that a high level of compatibility can allow restoration, contributing to particularly profound and lasting results. Many factors lead environments to have poor compatibility, including:

- *Unfamiliar places*: as previously noted, an unfamiliar environment requires the activation of multiple cognitive maps simultaneously, thus spending a greater amount of energy to understand the new environment.
- *Inadequate motivation*: an individual with little or different motivation, who finds no correspondence in the environment they are in, may perform the same activity but without experiencing the same restorative effects as another individual with greater motivation or in line with the characteristics of the environment.
- *Novelty of the task*: an unexpected task involves acquiring new skills, which do not allow for complete recovery. If an individual performs a task they intended to do, then they have prepared adequately for the task in question; however, if they are presented with a task or activity that is not exactly what they had anticipated doing, an incompatibility occurs.

In addition to the previously mentioned factors, Kaplan (2001) identifies six additional "incompatibility factors" that can inhibit the restorative experience:

Distraction: many environments are described as relaxing or distracting depending on the presence or absence of fascinating stimuli. However, a highly distracting place does not allow for the clear acquisition of information due to numerous stimuli and, consequently, increases the likelihood of fatigue from directed attention.

Information Deficit: an information deficit is experienced when an individual does not have adequate information about the environment, consequently needing to increase attention to find solutions useful for accomplishing the task.

- Danger: an environment perceived as dangerous by an individual evokes high levels of vigilance, with a consequent increase in sustained attention, which does not allow for cognitive recovery.
- Duty: to experience a sense of well-being, an individual should not feel drawn to the environment out of a sense of duty or responsibility but should be attracted by a desire for pleasure and recovery.
- Deception: a rejuvenating environment requires a match between the task the individual is performing and their thoughts about that task. A discrepancy between the two does not lead to effective recovery from mental fatigue.
- Difficulty: in the face of a difficult situation, the lack of preparation or competence leads the individual to have to use multiple and complex cognitive maps, which do not allow relaxation; the purpose of the experience is to restore depleted cognitive resources, not to utilize new ones.

These factors underscore the importance of ensuring environments are conducive to restoration by minimizing these elements of incompatibility. Understanding and managing these factors can greatly enhance the effectiveness of natural and designed restorative environments.

4.3   Virtual Reality: A New Frontier for Psychological Restoration and Well-Being

VR is emerging as a revolutionary technology capable of transporting us to different worlds without leaving the comfort of our homes. Originally conceived for entertainment and gaming, VR has found extremely varied applications, extending its use to sectors such as education, medicine, and personal well-being. Thanks to the increasingly widespread distribution of virtual reality systems, it is now possible to leverage these immersive environments to provide significant support in many daily activities.

Particularly interesting is the use of VR to create restorative experiences, especially for those seeking relief from work-related stress and fatigue, as well as for those physically unable to move. In contemporary society, characterized by hectic rhythms and increasing daily commitments, it is increasingly difficult to find the time and resources to physically move to ideal destinations for relaxation and energy recovery. This context has fueled interest in alternative solutions that can provide relief and well-being without the need to travel. VR presents itself as a promising technology in this field, offering the possibility of immersive experiences that allow users to "move without moving." Although a substitute for a real experience, VR represents a valid option for those who cannot afford the luxury of physical movement, still ensuring significant benefits for psychophysical well-being.

Recent studies have demonstrated that virtual reality can be effectively used to induce states of relaxation and improve psychological well-being. For example, a study conducted by Khirallah Abd El Fatah et al. (2024) exemplifies the therapeutic potential of VR, showing how the use of immersive reminiscence therapy, mediated by VR, can significantly improve the cognitive functions and psychological well-being of elderly individuals in care facilities. Their study, a randomized controlled trial, indeed compared immersive virtual reminiscence with traditional reminiscence therapy, highlighting statistically significant improvements in cognitive functions and psychological well-being of participants subjected to VR compared to the control group (Khirallah Abd El Fatah et al., 2024). This research demonstrates how VR can create an immersive and personalized experience that promotes mental well-being and reduces stress.

Among the various applications, the therapeutic uses of VR stand out, extending even to the treatment of neurodegenerative disorders, such as dementia. A relevant example of such application is the mixed pilot study conducted by Moyle et al. (2018), which examined the effectiveness of immersive exposure in a virtual forest to which patients suffering from dementia were subjected. In this study, the immersive VR environment was used to assess the impact on engagement, apathy, and mood states of the participants. The results demonstrated that VR has noticeable positive effects: during the experience in the virtual forest, patients showed an increase in pleasure and a higher level of alertness, although some also showed an increase in fear/anxiety. However, interviews conducted with staff, residents, and their families revealed a generally positive perception of the VR experience, highlighting how such technologies can offer significant emotional and cognitive benefits, improving interaction and quality of life of patients (Moyle et al., 2018).

In another study, conducted by Argüero-Fonseca et al. (2023b), the effects of an environmental psychological restoration protocol through VR were examined, aimed at countering demotivation in children, thus targeting a completely different demographic compared to the previous studies mentioned. Specifically, the study involved children aged between

8 and 11 years, subjecting them to VR sessions that simulated natural environments. The results indicated a significant decrease in signs of demotivation post-treatment, suggesting that exposure to nature, even if virtual, can have tangible benefits on children's motivation (Argüero-Fonseca et al., 2023).

VR not only allows individuals to "travel" virtually to otherwise inaccessible places but also confirms itself as a valuable tool for recovery from fatigue and stress, especially after intense workdays. Additionally, VR's ability to simulate naturalistic environments and situations that would otherwise be inaccessible positions it as a valuable tool for those who, after a day of work, lack the opportunity or time to go outdoors, offering a valid and restful alternative to the traditional walk in the green. Thus, VR not only expands the possibilities for relaxation and distraction but also actively contributes to recovery from fatigue and stress, supporting the overall well-being of the individual.

The application of VR as a relaxation and recovery tool after work fits perfectly into this vision. Using VR headsets, individuals can "detach" from their usual environment and immerse themselves in relaxing scenarios that promote recovery from daily stress. This practice can be particularly useful for those living in dense urban contexts, where opportunities for direct interaction with nature are limited.

The effectiveness of virtual reality in these applications is also supported by the ability of these environments to stimulate the mind and body in ways that other digital media cannot. The immersive nature of VR ensures that users can not only see but "feel" virtual environments, which evidently amplifies the benefits for psychological well-being.

4.4   CASE STUDY 3: EXPOSURE TO NATURAL ENVIRONMENTS IN VR

Introduction. Contemporary scientific research has recently observed an increased interest in exploring the intrinsic connection between the environment and mental well-being, an area of growing relevance in an era characterized by a hectic lifestyle and pervasive urbanization. This context has provided fertile ground for initiating a series of investigations, among which a recent study conducted in Italy by the research team of the Psychology Department of the Salesian University Institute of Venice-Mestre (Marchioro et al., 2023) is presented here as an example. The study is epistemologically situated within the theoretical framework of *Environmental Psychological Restoration—EPR* (Hartig et al., 1991; Kaplan & Kaplan, 1989; Martínez-Soto & González-Santos, 2020), which focuses on analyzing the process by which interaction with natural environments or green spaces facilitates the regeneration of individuals' cognitive and emotional resources. These resources can significantly diminish following periods of stress or mental fatigue (Kaplan & Talbot, 1983; Kaplan & Kaplan, 1989). The construct, in turn, is based on the *Attention Restoration Theory—ART*, developed by Rachel Kaplan and Stephen Kaplan (1989), Stephen Kaplan (1995), which postulates that certain environments (specifically natural ones) have the potential to positively influence our attention capacity, often compromised by the continuous demands and stress of everyday life (Kaplan, 1995). Notably, ART is therefore founded on the premise that it is precisely natural environments, thanks to their intrinsic characteristics, that facilitate this recovery, also aiding the reestablishment of the psychological balance that is often severely tested by the numerous demands occurring in urban or work environments.

Recent scientific findings have substantially confirmed the thesis that interaction with natural environments, including mere exposure to visual representations of landscapes, has a beneficial effect in restoring focused attention capacity and reducing stress levels, thereby bringing a qualitative improvement to the overall psychological well-being of individuals (Berto, 2005). However, the intensification of urbanization today has significantly reduced the possibilities for direct contact with natural spaces, which are becoming increasingly scarce (Hartig et al., 2014), relegating many individuals to spend most of their time in enclosed and artificial spaces, far from the vitality and regenerative potential of nature, which has now become a temporary refuge primarily during leisure time on weekends.

4.4.1 Designing Restorative Virtual Environments: Conceptualization and Operationalization in Virtual Reality

In this context, VR emerges as a significant technological innovation, offering the ability to recreate natural experiences that can elicit effects comparable to those achieved through direct contact with the natural environment (Valtchanov & Ellard, 2015). However, to recreate a virtual environment with these characteristics, it was necessary to precisely define the concept of a "restorative environment," starting from a generic definition and then narrowing the semantic field to facilitate a more precise application of the construct in the context of VR. The literature supports that restorative environments are those places that have the potential to foster the renewal of mental resources, which tend to deplete more quickly when we are in environments unsuitable for the activities we intend to perform (Kaplan & Kaplan, 1989; Kaplan, 1995; Hartig et al., 1997).

Operationalization Criteria for VR Environments. Considering these theoretical premises, the design and evaluation of restorative virtual environments referred to the Perceived Restorativeness Scale (PRS), developed and subsequently refined by Hartig and colleagues (Hartig et al., 1991, 1997). The PRS, in its brief form, was then used to assess the reliability and validity of the restorative quality of various environments, serving as a criterion for selecting and refining virtual scenarios in terms of restorative potential. This tool enabled the identification of properties that would maximize the restorative potential of environments in VR, thus guiding researchers in choosing among different design alternatives and in the subsequent optimization of environments for their definitive virtual implementation.

4.4.2   Methods

The research group from Venice-Mestre leveraged virtual reality (VR) as an alternative means to explore the regenerative properties of nature, aiming to overcome the logistical barriers that would have inevitably complicated direct observation in authentic natural environments.

4.4.2.1   Objectives and Goals of the Study

The study design was structured around a dual investigative purpose: on one side, it aimed to examine the effectiveness of images with high restorative potential, mediated by VR, in mitigating cognitive fatigue; on the other, the intent was to determine if the simulated experience could match, in its positive outcomes, the experience of immersion in real natural environments.

The primary goal was to explore the effect of exposure to restorative environments recreated in VR in mitigating cognitive fatigue (Mental Fatigue) in its acute phase, understood as a temporary decrement in cognitive capacity following intense mental efforts (Boksem & Tops, 2008). Specifically, the study aimed to compare the effect of high-potential restorative images against those of low potential, to assess the efficacy of VR as a potential restorative tool, usable in contexts of isolation due to physical illnesses or adverse environmental circumstances.

4.4.2.2 Sample and Experimental Design

The sample consisted of 70 university students, aged between 19 and 25 years (M = 20.69, SD = 1.57). The exclusion criterion was an out-of-norm IQ value, to prevent potential distortions of the results.[1] To assess the effects of exposure to natural environments through VR on mental fatigue, a mixed factorial design was employed. This methodological approach allowed for the incorporation of the gender of the participants as an additional stratification variable, thus providing an extra level of statistical control and enhancing the robustness of the findings.

The experimental design was structured in four phases: initially, participants were introduced to VR; this was followed by three rounds of mental fatigue induction, through the administration of *Raven's Advanced Progressive Matrices (APM)*, delivered in three blocks of 16 items each, using the *odd-even method* to ensure a balanced difficulty level. Performance was evaluated using the same APM, utilized both as a tool for inducing cognitive fatigue and as a measure for assessing the restorative effect of the virtual scenarios. The intervals between APM sessions included exposure to virtual contexts differentiated by restorative potential and dimensionality (3D vs 2D).

Participants were randomly assigned to one of four experimental conditions (Fig. 4.1), to evaluate the restorative efficacy of high-potential images compared to those of low potential, while also considering the difference between immersion in interactive three-dimensional (3D) environments and the viewing of static two-dimensional (2D) images, resulting in 4 independent groups (each corresponding to a specific condition):

- Group 1 (G1): high-potential restorative images in 3D;
- Group 2 (G2): low-potential restorative images in 3D;
- Group 3 (G3): high-potential restorative images in 2D;
- Group 4 (G4): low-potential restorative images in 2D.

---

[1] IQ values were based on previous academic performance data, as certified by university admission tests (Marchioro, 2023).

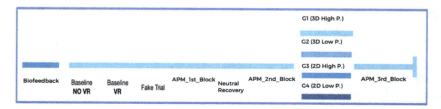

Fig. 4.1   Experimental procedure diagram

4.4.2.3   Hypotheses

Based on the theoretical premise and the objectives outlined, the study formulated the following operational hypotheses, divided into two main categories: the first concerning the efficacy of high restorative potential images, and the second concerning the efficacy of three-dimensional immersion compared to two-dimensional viewing.

- **Hypothesis 1**: It was predicted that exposure to high restorative potential images in VR would significantly reduce cognitive fatigue compared to exposure to low restorative potential images in VR.

  - *Hypothesis 1a*: Students exposed to high restorative potential images in 3D will show a greater reduction in cognitive fatigue compared to those exposed to low restorative potential images in 3D.
  - *Hypothesis 1b*: Subjects exposed to high restorative potential images in 2D will show a greater reduction in cognitive fatigue compared to those exposed to low restorative potential images in 2D.

- **Hypothesis 2**: Immersion in three-dimensional (3D) environments in VR will have a greater restorative effect on cognitive fatigue than viewing two-dimensional (2D) images.

  - *Hypothesis 2a:* Subjects exposed to high restorative potential images in 3D will show a greater reduction in cognitive fatigue compared to those exposed to high restorative potential images in 2D.

- *Hypothesis 2b:* Subjects exposed to low restorative potential images in 3D will show a greater reduction in cognitive fatigue compared to those exposed to low restorative potential images in 2D.

These hypotheses aim to explore the specific contributions of image restorativeness and dimensional immersion to cognitive fatigue alleviation within a VR context.

The intention was thus to explore the influence of the type and dimensionality of images on actual cognitive recovery capacity, as inferred from the performance achieved in each APM sub-test, assuming that better results indicated greater benefits from the recovery period.

Furthermore, for an in-depth monitoring of physiological parameters, Biofeedback was employed in each group throughout the entire administration phase: data were collected on heart rate, skin conductance, abdominal respiration, and frontal muscle tension. Measurements were conducted continuously, to accurately record the physiological fluctuations associated with processes of fatigue and potential cognitive recovery (Marchioro, 2023).

4.4.2.4 Study Procedure

Participants were welcomed by an operator into a room sufficiently spacious to allow them to freely use the Meta Quest 2 virtual reality headsets. They were then introduced to VR through the *First Step for Quest 2* application (Fig. 4.2), which, via a guiding voice, enabled them to become familiar with the controls and buttons of the controllers. This setting was designed both to minimize the sensation of disorientation that novice subjects might experience with virtual reality and VR headsets, and to acclimate them to this new experience, thus averting common risks associated with VR experiences, such as motion sickness.

After the initial setting phase, which lasted about 8 minutes, the participants were escorted to a second room where the actual test was administered. This involved inducing mental fatigue through the administration of the first block (Fig. 4.3) of the APM, comprising 16 items selected using the odd-even method to ensure a balanced level of difficulty.

At the end of the test, participants were exposed to a neutral virtual environment, in this case, an empty room with a gray wall, and were asked to do nothing for about two minutes. This allowed participants to recover some of their mental fatigue before responding to the second block of the Advanced Raven's Matrices (APM), which also comprised 16 items.

Fig. 4.2 The subject's avatar within the First Step for Quest 2 application holding a Ping Pong paddle

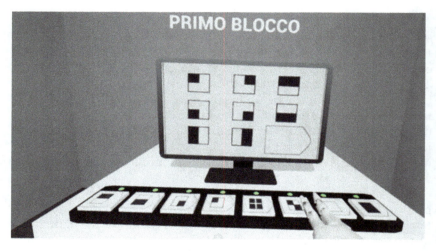

Fig. 4.3 Item from the First Block of Raven's Matrices in VR

Once the second set of APM was completed, individuals were exposed to different virtual contexts according to their group. Figure 4.4 illustrates the images corresponding to the two different types of scenarios.

Fig. 4.4   Virtual environment with low restorative potential (left) vs virtual environment with high restorative potential

4.4.2.5   Clarifications on the Use of Biofeedback (BF)

Among the tools available, Biofeedback (BF) proved necessary to monitor and record various physiological parameters of participants without providing video or audio feedback. This enabled researchers to obtain a real-time biological snapshot of each subject, record it, and compare it across different experimental settings. Specifically, the *Procomp Infiniti Biofeedback* system (8 Channels) from *Thought Technology*.

BF allowed for the recording of biophysical signals and monitoring the following physiological parameters:

– Muscle tension or electromyography (EMG).
– Thoracic and abdominal respiration.
– Cardiac indices (Heart Rate HR - Heart Rate Variability HRV).
– Skin conductance (SC) or sweating.

For recording electromyography (EMG), which measures muscle tension, the MyoScan-Pro™ EMG sensor with disposable adhesive sponge electrodes (30 nm diameter) was utilized, attached to the skin using a solid gel. Two bands were used for respiration: one for thoracic respiration and another for abdominal respiration, each equipped with the SA9311M respiration sensor. The SA9308M sensor was used to monitor cardiac pulse volume indices, while the SA9309M sensor was used for skin conductance. The measured signals were transmitted to a laptop, which ran specific software, *Biograph Infiniti*, for the visualization and recording of the signals.

Once the recording was completed, data cleansing was conducted to eliminate any potential artifacts using specific software.

4.4.3   Data Analysis and Results

The output from the mixed model for the analysis of variance (Mixed Model ANOVA), shown in Table 4.1, provides a scientifically robust illustration of the influence of experimental conditions on cognitive performance, measured through the APM test. The p-values for the "within-subjects" factor and its interactions were calculated using the Greenhouse-Geisser correction. The analysis of variance (ANOVA), conducted with a 95% confidence level, indeed revealed that variations in experimental conditions had a significant impact on participants' scores across different blocks of the APM test [$F_{(3, 65)}$ = 3.95, p = 0.012]. This implies that the various conditions imposed during the experiment—presumably related to variations in the virtual environment presented to the participants—significantly altered their ability to tackle the cognitive challenges posed by the test. The lack of statistical significance related to the gender of the participants suggests that performance differences in the APM test were not influenced by biological sex [$F_{(1, 65)}$ = 0.23, p = 0.634]. In other words, men and women reacted similarly to the experimental conditions in terms of cognitive fatigue and attention capacity.

The significant effect related to the interaction between the "within-subjects" factor and treatment (linked to the four different conditions) suggests that the experimental conditions influenced the scores in the various blocks of the APM test differently [$F_{(2, 130)}$ = 18.51, p < 0.001].

This indicates that some types of virtual environmental exposure were more effective than others in mitigating cognitive fatigue. For a more in-depth analysis, post hoc comparisons were performed to examine in more

Table 4.1   Mixed Model ANOVA results on cognitive performance as measured by APM test

Table. Source	df	SS	MS	F	p	η_p^2
Between-Subjects						
Condition	3	56.51	18.84	3.95	0.012	0.15
Gender	1	1.09	1.09	0.23	0.634	0.004
Residuals	65	310.07	4.77			
Within-Subjects						
Within Factor	2	24.73	12.37	18.51	<0.001	0.22
Condition: Within Factor	6	125.08	20.85	31.20	<0.001	0.59
Gender: Within Factor	2	1.59	0.80	1.19	0.299	0.02
Residuals	130	86.85	0.67			

detail the differences between the scores obtained by subjects in the various experimental conditions and results across APM test blocks. To make these comparisons, the Tukey method was adopted, with a significance level α of 0.05, to test the differences in the estimated marginal means for each combination of between-subject and within-subject effects. Regarding the between-subject effects, it was found that, in the category of 2D images with low restorative potential, the score of the first APM block was significantly higher than that of the third block ($p < 0.05$), and the score of the second block also significantly exceeded that of the third block ($p < 0.01$). In the category of 3D images with high restorative potential, the score of the first block was significantly higher than that of the second block ($p < 0.05$), while the score of the third block significantly exceeded both the first and second block scores ($p < 0.001$). For the 3D images with low restorative potential, the results were similar: the score of the first block was higher than that of the subsequent blocks ($p < 0.05$).

The results of the post hoc comparisons provide a clearer picture of the impact of the different experimental conditions on cognitive performance, offering further confirmation that the visual context, manipulated through restorative images, can play a significant role in cognitive performance as measured by APM tests. In particular, scenarios with high restorative potential three-dimensional images were the most effective for cognitive recovery, as evidenced by the improvement in scores following their presentation. For images with low restorative potential, it was observed that participants performed better in the early test blocks compared to the last one, suggesting that the impact of these images on cognitive performance evidently resulted in a decline in performance. In summary, the detailed analysis of post hoc results suggests that the visual context and restorative quality of images have a tangible impact on cognitive performance. These variations can be attributed to both the type of image and the sequence of task presentation to subjects. This provides significant insights for further studies on the therapeutic and enhancement potential of restorative images, emphasizing the essentiality of the environment as an influencing factor in the regulation of cognitive functions.

4.4.4 Limitations and Future Developments

The merit of the presented research undoubtedly lies in highlighting the crucial importance of well-designed virtual environments in promoting cognitive restoration, in line with Kaplan's (1989) *Attention Restoration*

Theory (ART) and Ulrich's (1983) studies on the psychophysiological response to natural environments (Ulrich, 1983). Kaplan (2001) also posits that much depends on individual dispositions: regardless of the environment's degree of fascination, if a person is not sufficiently receptive to the environment, they will not benefit from it.

The effectiveness of virtual reality, as demonstrated by the case study, in replicating the restorative qualities of natural environments opens innovative perspectives for potential interventions in the psychological field (Argüero-Fonseca et al., 2023), particularly in contexts where direct interaction with nature is impractical or limited. This is a widely corroborated theme in existing literature, which recognizes the therapeutic value of natural environments for mental health (Hartig et al., 2014; Capaldi et al., 2014).

Moreover, this research integrates into the broader context of Integral Ecology, offering a lens through which to observe the interconnection between humans and the environment, aiming to promote a deep understanding of the reciprocal influence between ecological and psychological health through practices that respect and regenerate both aspects (Marchioro, 2023).

It is also worth noting that, following the indications of Valtchanov and Ellard (2015), immersive technologies can serve as a means to counter the growing disconnection from the natural environment, a consequence of urbanization. Exploring the potential of such technological tools for enriching everyday life, strengthening mental health, and promoting a more sustainable ecological balance is certainly an area of considerable interest.

However, the unresolved issue of the accessibility and effective integration of these technologies into clinical practice and daily routines persists. Future research should, therefore, focus on identifying cost-effective and widespread strategies to incorporate virtual reality into the social fabric as part of a holistic approach to mental well-being.

Finally, this discussion would be incomplete without a critical reflection on the ethics of using virtual reality. It is imperative that future investigations consider the potential risks associated with immersiveness and technological dependence, as well as the implications of equitable distribution of technological resources, to ensure that the use of virtual reality is guided by ethical principles that promote the common good and social justice.

4.5 Concluding Remarks

This work has sought to explore the convergence between VR technology and environmental psychology, drawing upon the pioneering work of Gifford (2014), who examined the symbiosis between the human psyche and the environment.

This disciplinary intersection is strengthened by the convergence of various research strands that have addressed the therapeutic effects of natural environments on cognition and emotional well-being (Kaplan & Kaplan, 1989; Ulrich, 1983), as well as sustainable practices and policies that promote environmental health (Hartig et al., 2014). Additionally, studies on the regenerative potentials of nature, even when mediated by immersive technologies like virtual reality (Scopelliti & Giuliani, 2004; Berto, 2005; Marchioro et al., 2023; Marchioro, 2023; Argüero-Fonseca et al., 2023), confirm the efficacy of such environments, even when simulated, in promoting cognitive recovery and reducing stress, in line with Kaplan's (1989) assertions on the Attention Restoration Theory (ART).

Today more than ever, we feel the need to adopt innovative strategies to maintain and strengthen our connection with the natural environment, which may be lost or weakened by the distance (both physical and spiritual) that separates most people from nature.

In the context of an increasingly urbanized and technology-dependent society, VR, although it may be considered merely a surrogate for reality, still offers an alternative access to the regenerative benefits of nature. A line of research in this direction not only corroborates an ecological view but could also suggest how technology, if used ethically, can be employed to enrich our connection with the environment.

In this vision, technology and ecology not only intersect but also enhance each other, offering a richer and more complex interpretation of the relationship between the human mind and its vital context, represented by the Environment.

References

Afari, N., & Buchwald, D. (2003). Chronic fatigue syndrome: A review. *American Journal of Psychiatry, 160*(2), 221–236. https://doi.org/10.1176/appi.ajp.160.2.221

Argüero-Fonseca, A., Martínez Soto, J., Barrios Payán, F. A., Villaseñor Cabrera, T. D., Reyes Huerta, H. E., González Santos, L., Aguirre Ojeda, D. P., Pérez

Pimienta, D., Reynoso González, O. U., & Marchioro, D. (2023a). Effects of an n-back task on indicators of perceived cognitive fatigue and fatigability in healthy adults. *Acta Biomed, 94*(6), 1–12. https://doi.org/10.23750/abm.v94i1.15649

Argüero-Fonseca, A., Martínez-Soto, J., Aguirre-Ojeda, D. P., Pérez-Pimienta, D., & Marchioro, D. M. (2023b). Effects of a protocol of environmental psychological restoration with virtual reality on indicators of demotivation in children. *Journal of Population Therapeutics & Clinical Pharmacology*, 821–828. https://doi.org/10.53555/jptcp.v30i18.3182

Bafna, T., Bækgaard, P., & Hansen, J. (2021). Mental fatigue prediction during eye-typing. *PLoS One, 16*(2), 1–17. https://doi.org/10.1371/journal.pone.0246739

Berto, R. (2005). Exposure to restorative environments helps restore attentional capacity. *Journal of Environmental Psychology, 25*(3), 249–259. https://doi.org/10.1016/j.jenvp.2005.07.001

Boksem, M. A., & Tops, M. (2008). Mental fatigue: Costs and benefits. *Brain Research Reviews, 59*(1), 125–139. https://doi.org/10.1016/j.brainresrev.2008.07.001

Bültmann, U., Kant, I., Kasl, S. V., Beurskens, A. J., & van den Brandt, P. A. (2002). Fatigue and psychological distress in the working population: Psychometrics, prevalence, and correlates. *Journal of Psychosomatic Research, 52*(6), 445–452. https://doi.org/10.1016/s0022-3999(01)00228-8

Capaldi, C. A., Dopko, R. L., & Zelenski, J. M. (2014). The relationship between nature connectedness and happiness: A meta-analysis. *Frontiers in Psychology, 5*, 1–15. https://doi.org/10.3389/fpsyg.2014.00976

Cornuz, J., Guessous, I., & Favrat, B. (2006). Fatigue: A practical approach to diagnosis in primary care. *Canadian Medical Association Journal, 174*(6), 765–767. https://doi.org/10.1503/cmaj.1031153

David, A., Pelosi, A., McDonald, E., Stephens, D., Ledger, D., Rathbone, R., & Mann, A. (1990). Tired, weak, or in need of rest: Fatigue among general practice attenders. *The BMJ, 301*(6762), 1199–1202. https://doi.org/10.1136/bmj.301.6762.1199

Doherty, T. J., & Clayton, S. (2011). The psychological impacts of global climate change. *American Psychologist, 66*(4), 265–276. https://doi.org/10.1037/a0023141

Faber, L. G., Maurits, N. M., & Lorist, M. M. (2012). Mental fatigue affects visual selective attention. *PLoS One, 7*(10), 1–10. https://doi.org/10.1371/journal.pone.0048073

Fritze, J. G., Blashki, G. A., Burke, S., & Wiseman, J. (2008). Hope, despair and transformation: Climate change and the promotion of mental health and well-being. *International Journal of Mental Health Systems, 2*(13), 1–10. https://doi.org/10.1186/1752-4458-2-13

Fuhrer, R., & Wessely, S. (1995). The epidemiology of fatigue and depression: A French primary-care study. *Psychological Medicine, 25*(5), 895–905. https://doi.org/10.1017/s0033291700037387

Galland-Decker, C., Marques-Vidal, P., & Vollenweider, P. (2019). Prevalence and factors associated with fatigue in the Lausanne middle-aged population: A population-based, cross-sectional survey. *BMJ Open, 9*(8), 1–10. https://doi.org/10.1136/bmjopen-2018-027070

Gifford, R. (2014). Environmental psychology matters. *Annual Review of Psychology, 65*, 541–579. https://doi.org/10.1146/annurev-psych-010213-115048

Guifang, M., Minpeng, X., Chuncui, Z., Feng, H., Hongzhi, Q., & Dong, M. (2017). A specific EEG network structure related to mental fatigue induced by N-back. *Chinese Journal of Biomedical Engineering, 36*(2), 143–149. https://doi.org/10.3969/j.issn.0258-8021.2017.02.003

Harrington, M. (2012). Neurobiological studies of fatigue. *Progress in Neurobiology, 99*(2), 93–105. https://doi.org/10.1016/j.pneurobio.2012.07.004

Hartig, T., Korpela, K., Evans, G. W., & Gärling, T. (1997). A measure of restorative quality in environments. *Scandinavian Housing and Planning Research, 14*(4), 175–194. https://doi.org/10.1080/02815739708730435

Hartig, T., Mang, M., & Evans, G. W. (1991). Restorative effects of natural environment experiences. *Environment and Behavior, 23*(1), 3–26. https://doi.org/10.1177/0013916591231001

Hartig, T., Mitchell, R., de Vries, S., & Frumkin, H. (2014). Nature and health. *Annual Review of Public Health, 35*(1), 207–228. https://doi.org/10.1146/annurev-publhealth-032013-182443

Herzog, T., Maguire, P., & Nebel, M. B. (2003). Assessing the restorative components of environments. *Journal of Environmental Psychology, 23*(2), 159–170. https://doi.org/10.1016/S0272-4944(02)00113-5

Hopstaken, J. F., van der Linden, D., Bakker, A. B., & Kompier, M. A. (2015). A multifaceted investigation of the link between mental fatigue and task disengagement. *Psychophysiology, 52*(3), 305–331. https://doi.org/10.1111/psyp.12339

Jaeggi, S. M., Buschkuehl, M., Perrig, W. J., & Meier, B. (2010). The concurrent validity of the N-back task as a working memory measure. *Memory, 18*(4), 394–412. https://doi.org/10.1080/09658211003702171

James, W. (1892). *Psychology: The briefer course.* H. Holt & Co.

Kaplan, R., & Kaplan, S. (1989). *The experience of nature: A psychological perspective.* Cambridge University Press.

Kaplan, S. (1995). The restorative benefits of nature: Toward an integrative framework. *Journal of Environmental Psychology, 15*(3), 169–182. https://doi.org/10.1016/0272-4944(95)90001-2

Kaplan, S. (2001). Meditation, restoration, and the management of mental fatigue. *Environment and Behavior, 33*(4), 480–506. https://doi.org/10.1177/00139160121973106

Kaplan, S., & Talbot, J. F. (1983). Psychological benefits of a wilderness experience. In I. I. Altman & J. F. Wohlwill (Eds.), *Behavior and the natural environment* (Vol. 6, pp. 163–203). Springer.

Kellert, S. R., & Wilson, E. O. (1993). *The biophilia hypothesis.* Island Press.

Khirallah Abd El Fatah, N., Abdelwahab Khedr, M., Alshammari, M., & Mabrouk Abdelaziz Elgarhy, S. (2024). Effect of immersive virtual reality reminiscence versus traditional reminiscence therapy on cognitive function and psychological well-being among older adults in assisted living facilities: A randomized controlled trial. *Geriatric Nursing, 55,* 191–203. https://doi.org/10.1016/j.gerinurse.2023.11.010

Lorist, M. M., Boksem, M. A., & Ridderinkhof, K. R. (2005). Impaired cognitive control and reduced cingulate activity during mental fatigue. *Cognitive Brain Research, 24*(2), 199–205. https://doi.org/10.1016/j.cogbrainres.2005.01.018

Marchioro, D. M. (2023). La Psicologia nell'era dell'Ecologia integrale: un percorso di sinergie e complementarità. *IUSVEducation, 22,* 48–65. https://www.iusveducation.it/la-psicologia-nellera-dellecologia-integrale-un-percorso-di-sinergie-e-complementarita/

Marchioro, D. M., Argüero-Fonseca, A., Bounous, M., & Benatti, F. (2023). Effectiveness of images with high-potential restorative in virtual reality to reduce acute cognitive fatigue in undergraduate students. In *18th European congress of psychology - ECP. Psychology: Uniting communities for a sustainable world* (p. 274). European Federation of Psychologists' Associations.

Martínez-Soto, J., & González-Santos, L. (2020). Restauración psicológica afectiva a partir de la exposición mediada del ambiente [Affective psychological restoration through mediated exposure to the environment]. *PsyEcology, 11*(3), 289–318. https://doi.org/10.1080/21711976.2020.1730133

Mota, D. D., & Pimenta, C. A. (2006). Self-report instruments for fatigue assessment: A systematic review. *Research and Theory for Nursing Practice, 20*(1), 49–78. https://doi.org/10.1891/rtnp.20.1.49

Moyle, W., Jones, C., Dwan, T., & Petrovich, T. (2018). Effectiveness of a virtual reality Forest on people with dementia: A mixed methods pilot study. *The Gerontologist, 58*(3), 478–487. https://doi.org/10.1093/geront/gnw270

Ohly, H., White, M. P., Wheeler, B. W., Bethel, A., Ukoumunne, O. C., Nikolaou, V., & Garside, R. (2016). Attention restoration theory: A systematic review of the attention restoration potential of exposure to natural environments. *Journal of Toxicology and Environmental Health. Part B, Critical Reviews, 19*(7), 305–343. https://doi.org/10.1080/10937404.2016.1196155

Pawlikowska, T., Chalder, T., Hirsch, S. R., Wallace, P., Wright, D., & Wessely, S. (1994). Population based study of fatigue and psychological distress. *The BMJ, 308*(6931), 763–766. https://doi.org/10.1136/bmj.308.6931.763

Ragsdale, J. M., Beehr, T. A., Grebner, S., & Kyunghee, H. (2011). An integrated model of weekday stress and weekend recovery of students. *International Journal of Stress Management, 18*(2), 153–180. https://doi.org/10.1037/a0023190

Raven, J. (2000). The Raven's progressive matrices: Change and stability over culture and time. *Cognitive Psychology, 41*(1), 1–48. https://doi.org/10.1006/cogp.1999.0735

Scopelliti, M., & Giuliani, M. V. (2004). Choosing restorative environments across the lifespan: A matter of place experience. *Journal of Environmental Psychology, 24*(4), 423–437. https://doi.org/10.1016/j.jenvp.2004.11.002

Smith, M. R., Chai, R., Nguyen, H. T., Marcora, S. M., & Coutts, A. J. (2019). Comparing the effects of three cognitive tasks on indicators of mental fatigue. *The Journal of Psychology, 153*(8), 759–783. https://doi.org/10.108 0/00223980.2019.1611530

Stokols, D. (1990). Instrumental and spiritual views of people-environment relations. *American Psychologist, 45*(5), 641–646. https://doi.org/10.1037/0003-066X.45.5.641

Ulrich, R. S. (1983). Aesthetic and affective response to natural environment. In I. Altman & J. F. Wohlwill (Eds.), *Human behavior and the natural environment* (pp. 85–125). Plenum.

Valtchanov, D., & Ellard, C. G. (2015). Cognitive and affective responses to natural scenes: Effects of low level visual properties on preference, cognitive load and eye-movements. *Journal of Environmental Psychology, 43*, 184–195. https://doi.org/10.1016/j.jenvp.2015.07.001

CHAPTER 5

Enhancing Motivation in Children with Virtual Reality

Abstract This chapter explores the transformative use of virtual reality (VR) to enhance motivation in children, addressing challenges in contexts such as education and therapy. It focuses on Environmental Psychological Restoration (EPR) protocols, which use VR to recreate natural environments, fostering psychological renewal and combating demotivation. A case study titled "Effects of a Protocol of Environmental Psychological Restoration with Virtual Reality on Indicators of Demotivation in Children" provides valuable insights into how VR can revitalize children's enthusiasm and participation. This study highlights VR's ability to adapt to diverse individual needs and learning styles, offering personalized and immersive experiences. The chapter underscores VR's potential to inspire and engage children, envisioning a future where this technology becomes a fundamental tool in education and therapy. By presenting VR as a means to explore, learn, and grow, it argues for its capacity to revolutionize the approach to learning and personal development in the younger generation. In summary, this chapter presents a compelling argument for the power of VR to transform children's motivation and engagement in their learning and personal development processes.

Keywords Virtual reality • Demotivation in children • Education • Psychotherapy • Environmental psychological restoration (EPR)

© The Author(s), under exclusive license to Springer Nature   119
Switzerland AG 2024
D. M. Marchioro et al., *Virtual Reality: Unlocking Emotions and Cognitive Marvels*, Palgrave Studies in Cyberpsychology,
https://doi.org/10.1007/978-3-031-68196-7_5

5.1   INTRODUCTION

Motivation is a critical factor in the educational and therapeutic development of children. High levels of motivation can significantly enhance learning outcomes, participation, and overall psychological well-being (Ryan & Deci, 2000). Conversely, demotivation can lead to disengagement, poor academic performance, and a lack of enthusiasm in therapy sessions (Fredricks et al., 2004). Addressing these issues, particularly in today's rapidly evolving technological landscape, is crucial. Virtual Reality (VR) offers a promising solution, providing immersive and engaging experiences that can potentially revolutionize traditional approaches to education and therapy (Parsons & Cobb, 2011).

In recent years, VR technology has advanced significantly, becoming more accessible and versatile (Freina & Ott, 2015). Its application in educational and therapeutic settings has opened new avenues for enhancing motivation among children. This chapter delves into the concept of Environmental Psychological Restoration (EPR) protocols, which utilize VR to recreate natural environments, promoting psychological renewal and combating demotivation (Kjellgren & Buhrkall, 2010). By examining the effects of VR-based EPR on children's motivation, we aim to highlight the transformative potential of this technology in fostering enthusiasm and participation (Riva et al., 2007).

Virtual Reality (VR) has emerged as a significant tool in enhancing motivation among children, often surpassing traditional methods in several key areas. Traditional methods, while effective, often struggle to maintain the engagement and interest of young learners, especially in the context of today's technology-driven world. VR, on the other hand, offers immersive and interactive experiences that can captivate children's attention and foster a deeper level of engagement (Parsons & Cobb, 2011). One major advantage of VR is its ability to create highly engaging and dynamic learning environments. These environments can simulate real-world scenarios and provide immediate feedback, making learning more interactive and enjoyable (Freina & Ott, 2015). This interactive nature of VR can lead to increased motivation as children are more likely to remain engaged with the material and participate actively in learning activities (Merchant et al., 2014).

Moreover, VR can be tailored to individual learning styles and needs, offering personalized learning experiences that can adapt to the pace and preferences of each child (Makransky & Lilleholt, 2018). This customization is often difficult to achieve with traditional methods, which tend to follow a one-size-fits-all approach. By catering to individual needs, VR can help in maintaining high levels of motivation and preventing the disengagement that can occur when children struggle to keep up with the standard curriculum.

In therapeutic settings, VR has been shown to effectively engage children by creating therapeutic environments that are both safe and stimulating. For instance, VR can be used to simulate social scenarios for children with autism, providing them with a controlled environment to practice social interactions without the stress of real-world consequences (Maskey et al., 2014). This approach not only enhances motivation but also helps in building confidence and reducing anxiety. Comparatively, traditional therapeutic methods may lack the same level of engagement and adaptability, often relying on repetitive tasks that can lead to boredom and decreased motivation over time. VR's ability to offer varied and interactive therapeutic exercises can keep children motivated and committed to their therapy programs (Lau et al., 2017).

This chapter aims to explore the transformative role of Virtual Reality (VR) in enhancing children's motivation through Environmental Psychological Restoration (EPR) protocols. The chapter is designed to provide a comprehensive overview of the theoretical foundations, practical implementations, and psychological benefits of VR in educational and therapeutic contexts. The structure of the chapter is outlined as follows:

Theoretical Foundations. The first section delves into the theoretical underpinnings that form the basis of this study. It begins with a detailed review of major motivational theories applicable to childhood. This review aims to provide a deep understanding of how motivation develops and operates in children. The discussion then moves to the application of VR in education and therapy, exploring how this technology has been utilized in these fields and its potential to significantly boost motivation. Finally, the concept of Environmental Psychological Restoration (EPR) is introduced. This part explains how EPR, facilitated through VR, can be relevant to enhancing motivation and psychological well-being in children.

Implementation of EPR Protocols with VR. The second section focuses on the practical aspects of implementing EPR protocols using VR. It begins with the design and development of these protocols, providing insights into how VR environments are created and tailored to be effective. Following this, the section discusses the simulation of natural environments in VR, emphasizing the importance of these settings in the restoration process. This part outlines the methods used to recreate these environments accurately and effectively. Finally, the expected psychological benefits of VR-based EPR, such as psychological renewal and rejuvenation, are detailed, highlighting the positive outcomes of these interventions.

Case Study: Effects of an EPR Protocol with VR on Indicators of Demotivation in Children. The third section presents a comprehensive case study that examines the effects of an EPR protocol with VR on indicators of demotivation in children. This part is divided into three subsections: methodology, results, and analysis and discussion. The methodology sub-section describes the experimental design, sample characteristics, and measurement instruments used in the study. The results sub-section presents the findings of the study, showcasing the impact of the EPR protocol on children's motivation levels. The analysis and discussion sub-section interprets these results within the context of existing literature, discussing the practical implications and potential for wider application.

Adaptability and Personalization of VR. The fourth section explores the adaptability and personalization capabilities of VR. It explains how VR experiences can be customized to meet the individual needs and preferences of children. This personalization is crucial for maximizing the effectiveness of VR interventions. Additionally, the section discusses how VR can cater to a diversity of learning styles, thereby enhancing motivation and engagement among children with different learning preferences.

Future Implications and Conclusions. The final section looks ahead to the future of VR in education and therapy. It envisions VR becoming a fundamental tool in these fields, offering innovative solutions to traditional challenges. This section also provides practical recommendations for implementing VR in educational and therapeutic settings, offering guidance on how to integrate this technology effectively. The chapter concludes with a summary of key points and reflections on the transformative potential of VR in enhancing children's motivation and overall development.

5.2   Theoretical Foundations

Key motivational theories, such as those by Ryan and Deci (2000), emphasize the importance of intrinsic and extrinsic factors in fostering motivation among children. VR has emerged as a powerful tool to enhance motivation through immersive and interactive experiences, significantly benefiting educational and therapeutic settings (Freina & Ott, 2015; Parsons & Cobb, 2011). VR's ability to create engaging learning environments and personalized therapeutic interventions has been shown to maintain high levels of engagement and adaptability (Makransky & Lilleholt, 2018; Maskey et al., 2014). EPR, when facilitated through VR, can simulate natural environments that promote psychological renewal and combat demotivation, thereby improving overall psychological well-being (Kjellgren & Buhrkall, 2010).

5.2.1   Theories of Motivation in Children

This section presents a review of the major motivational theories applicable to childhood. Understanding how motivation develops and operates in children is crucial for improving educational and therapeutic outcomes.

5.2.1.1 Self-Determination Theory

The Self-Determination Theory (SDT) developed by Ryan and Deci (2000) focuses on intrinsic and extrinsic motivation. According to this theory, intrinsic motivation, which is driven by interest and enjoyment of the activity itself, is essential for the well-being and performance of children. Ryan and Deci argue that intrinsic motivation flourishes when the basic psychological needs of autonomy, competence, and relatedness are satisfied. When children feel that they have control over their actions, believe in their abilities, and feel connected to others, they are more likely to be motivated and engaged in their activities (Ryan & Deci, 2000).

5.2.1.2 Expectancy-Value Theory

Expectancy-Value Theory, proposed by Eccles et al. (1983), suggests that a child's motivation is determined by their expectations of success and the value they place on success. This theory emphasizes that children are motivated to engage in tasks they find valuable and believe they can succeed in. The perceptions of competence and task value are influenced by social and cultural factors, including feedback from parents, teachers, and peers (Eccles et al., 1983).

5.2.1.3 Goal Orientation Theory

Goal Orientation Theory, developed by Dweck (1986), differentiates between mastery-oriented goals and performance-oriented goals. Mastery goals focus on learning and understanding, while performance goals are concerned with demonstrating ability relative to others. Children with mastery goals are more likely to be intrinsically motivated and show resilience in the face of challenges. In contrast, those with performance goals may be more extrinsically motivated and susceptible to anxiety and avoidance behaviors (Dweck, 1986).

5.2.1.4 Social Cognitive Theory

Bandura's Social Cognitive Theory (1986) highlights the role of self-efficacy in motivation. Self-efficacy refers to a child's belief in their ability to succeed in specific tasks. High self-efficacy can enhance motivation by increasing effort, persistence, and resilience. This theory also underscores the importance of observational learning, where children learn and become motivated by observing the behaviors and outcomes of others (Bandura, 1986).

By understanding these theories, educators and therapists can develop strategies to foster motivation in children, tailored to their individual needs and contexts. Integrating these theoretical perspectives can lead to more effective and engaging learning and therapeutic experiences.

Different motivational theories offer unique insights that can be applied to educational strategies to enhance children's learning experiences. For example, Self-Determination Theory (SDT) emphasizes the importance of fostering intrinsic motivation by satisfying children's needs for autonomy, competence, and relatedness (Ryan & Deci, 2000). Educators can design activities that provide choices, challenges, and opportunities for collaboration to meet these needs. Expectancy-Value Theory suggests that children are more motivated when they believe they can succeed and when they value the task (Eccles et al., 1983). Therefore, setting achievable goals and demonstrating the relevance of learning tasks can enhance motivation. Goal Orientation Theory advocates for promoting mastery goals over performance goals to encourage a focus on learning and personal improvement rather than competition (Dweck, 1986). Lastly, Social Cognitive Theory underscores the importance of building self-efficacy through positive feedback and role models (Bandura, 1986).

Intrinsic motivation refers to engaging in an activity for its own sake, driven by interest, enjoyment, and inherent satisfaction. In childhood learning, intrinsically motivated children are more likely to exhibit curiosity, persistence, and a desire to explore and understand new concepts. They tend to perform better academically and are more resilient in the face of challenges (Ryan & Deci, 2000). Extrinsic motivation, on the other hand, involves engaging in an activity due to external rewards or pressures, such as grades, praise, or avoidance of punishment. While extrinsic motivation can be effective in the short term, it may not sustain long-term engagement and can sometimes undermine intrinsic motivation if overemphasized (Deci et al., 1999). Effective educational strategies often balance both types of motivation, using extrinsic rewards to encourage initial engagement while fostering intrinsic motivation to maintain long-term interest and commitment.

Teachers and therapists can use motivational theories to create more engaging and effective learning and therapeutic environments. By applying Self-Determination Theory, they can design activities that promote autonomy, competence, and relatedness, such as offering choices in assignments, providing challenging yet attainable tasks, and encouraging cooperative learning (Ryan & Deci, 2000). Expectancy-Value Theory suggests that educators should set clear expectations for success and emphasize the value of learning activities to enhance motivation (Eccles et al., 1983). Goal Orientation Theory recommends focusing on mastery goals to encourage a growth mindset and resilience in children (Dweck, 1986). Social Cognitive Theory highlights the importance of building self-efficacy by providing positive feedback, modeling successful behaviors, and creating opportunities for observational learning (Bandura, 1986). By integrating these theories, teachers and therapists can develop holistic strategies that address the diverse motivational needs of children, leading to improved engagement and outcomes.

5.2.2   Virtual Reality in Education and Therapy

Virtual Reality (VR) has emerged as a powerful tool in both educational and therapeutic settings, offering immersive and interactive experiences that have the potential to significantly enhance motivation among children. This section discusses how VR has been utilized in these fields and its potential to increase motivation, supported by various theoretical perspectives and empirical studies.

5.2.2.1   VR in Education

VR in education provides a dynamic and engaging way to present educational content, making learning more interactive and enjoyable. The immersive nature of VR can simulate real-world scenarios, allowing students to explore and interact with environments that would otherwise be inaccessible. This aligns with constructivist theories of learning, which emphasize the importance of active, experiential learning (Dewey, 1938). For example, VR can enable students to take virtual field trips, conduct virtual lab experiments, and engage in historical reenactments, thereby enhancing their engagement and motivation to learn (Merchant et al., 2014).

Studies have shown that VR can improve various educational outcomes, including knowledge retention, conceptual understanding, and student engagement. Makransky and Lilleholt (2018) found that VR simulations in science education increased students' intrinsic motivation and perceived learning compared to traditional teaching methods. The study highlights that the immersive and interactive nature of VR can make learning more appealing and enjoyable, which in turn fosters a deeper engagement with the material (Makransky & Lilleholt, 2018).

5.2.2.2 VR in Therapy

In therapeutic settings, VR has been used to create controlled and safe environments for patients to undergo various therapeutic interventions. VR can simulate situations that are difficult to replicate in real life, such as social interactions for individuals with autism or exposure therapy for individuals with phobias. The ability to create customizable and repeatable scenarios makes VR a valuable tool in behavioral and psychological therapies (Riva et al., 2007).

For children, VR-based therapy can be particularly motivating due to its interactive and game-like nature. Maskey et al. (2014) demonstrated that VR environments could effectively reduce anxiety in children with autism spectrum disorder by providing a controlled setting for them to practice social interactions. The study showed that children were more willing to participate in VR-based therapy sessions, indicating higher levels of engagement and motivation (Maskey et al., 2014).

5.2.2.3 Potential to Increase Motivation

The potential of VR to increase motivation lies in its ability to provide engaging, interactive, and personalized experiences. According to the Self-Determination Theory (Ryan & Deci, 2000), motivation is enhanced when individuals experience autonomy, competence, and relatedness. VR can support these needs by offering choices within virtual environments, providing immediate feedback and challenges appropriate to the user's skill level, and enabling social interactions through collaborative virtual tasks.

Furthermore, Environmental Psychological Restoration (EPR) theory suggests that exposure to natural environments can restore attention and reduce mental fatigue (Kaplan & Kaplan, 1989). VR can simulate natural settings, offering children a virtual escape that can promote psychological renewal and enhance overall well-being, thereby combating demotivation (Kjellgren & Buhrkall, 2010).

5.2.3   Environmental Psychological Restoration (EPR)

Environmental Psychological Restoration (EPR) is a concept that refers to the ability of certain environments to promote psychological recovery and well-being. This concept is rooted in the Attention Restoration Theory developed by Kaplan and Kaplan (1989), which suggests that exposure to natural environments can help restore attention capacity and reduce mental fatigue. EPR is relevant to both motivation and psychological well-being, as restorative environments can enhance cognitive functioning, reduce stress, and improve overall mood.

5.2.3.1 Concept of EPR

The core idea of EPR is that natural environments possess restorative qualities that can replenish depleted cognitive resources. According to Kaplan and Kaplan, environments that are inherently fascinating provide a sense of being away, have extent (i.e., are rich and coherent), and offer compatibility with an individual's preferences and activities, are most effective in promoting restoration (Kaplan & Kaplan, 1989). Such environments allow individuals to recover from mental fatigue, leading to improved concentration, mood, and overall well-being.

According to the Attention Restoration Theory (ART) developed by Kaplan and Kaplan (1989), restorative environments have four key components:

Being Away: This refers to the sense of being removed from everyday stressors and routines, allowing the mind to relax.

Fascination: Environments that capture attention effortlessly (soft fascination), such as natural settings, help in restoring directed attention capacity.

Extent: The environment should be rich and coherent, offering a sense of scope and connectedness to immerse the individual.

Compatibility: The environment should be compatible with the individual's purposes and inclinations, allowing for a seamless interaction between the person and the setting (Kaplan & Kaplan, 1989).

5.2.3.2 Benefits of EPR

The psychological and physiological benefits of Environmental Psychological Restoration (EPR) in children are substantial:

Reduction in Stress and Anxiety: EPR can lower cortisol levels and reduce symptoms of stress and anxiety, creating a calming effect on the mind and body (Ulrich et al., 1991).

Improvement in Mood and Emotional Well-Being: Engaging with restorative environments enhances mood and reduces feelings of depression, promoting overall emotional well-being (Korpela et al., 2008).

Enhanced Cognitive Functioning: Restorative environments help replenish cognitive resources, improving attention, memory, and overall cognitive performance (Berto, 2005).

Increased Motivation and Engagement: By creating immersive and engaging experiences, VR-based EPR can boost motivation and engagement in both educational and therapeutic contexts (Makransky & Lilleholt, 2018).

Overall Psychological Well-Being: Regular exposure to restorative environments through VR can lead to a balanced mental state, reduced mental fatigue, and a general sense of well-being (Kjellgren & Buhrkall, 2010).

5.2.3.3 Application of EPR Through VR

Virtual Reality (VR) technology has made it possible to simulate natural restorative environments, providing the benefits of EPR even in urban or indoor settings. VR-based EPR can recreate various natural landscapes, offering immersive experiences that promote psychological restoration and combat demotivation. Kjellgren and Buhrkall (2010) found that simulated natural environments through VR can be as effective as real natural settings in promoting restoration and improving psychological well-being (Kjellgren & Buhrkall, 2010).

Virtual Reality (VR) can create effective restorative environments by replicating the key components of natural restorative settings:

Simulating Natural Environments: VR can generate realistic natural environments that provide the sense of being away and soft fascination, such as virtual forests, beaches, or mountains.

Immersive Experiences: VR offers a sense of extent by providing a 360-degree immersive experience that feels expansive and all-encompassing.

Interactive Elements: VR environments can be designed to be compatible with user preferences, allowing interaction with the virtual setting, such as walking through a forest or listening to the sounds of nature.

Customization: VR allows for personalized experiences tailored to individual needs, enhancing the compatibility component of ART (Makransky & Lilleholt, 2018).

5.3   IMPLEMENTATION OF EPR PROTOCOLS WITH VR

5.3.1   Development of EPR Protocols

The development of Environmental Psychological Restoration (EPR) protocols using Virtual Reality (VR) involves several key steps. Initially, the process starts with identifying the target population and their specific psychological needs. The design of the VR environment must align with the principles of the Attention Restoration Theory (ART) by Kaplan and Kaplan (1989), ensuring that the environments offer a sense of being away, fascination, extent, and compatibility (Kaplan & Kaplan, 1989).

The next phase involves creating detailed blueprints of the VR environments. This includes choosing appropriate natural settings such as forests, beaches, or mountains, which are known for their restorative qualities. The design process should incorporate input from psychologists, VR developers, and subject matter experts to ensure that the virtual environments are both realistic and therapeutically effective. User testing is also crucial, as it provides feedback on the usability and effectiveness of the VR environments in achieving the desired psychological outcomes (Gaggioli et al., 2015).

5.3.2   Simulation of Natural Environments

Recreating natural environments in VR is fundamental to the success of EPR protocols. The simulation process involves high-fidelity 3D modeling, realistic textures, and dynamic lighting to mimic the natural world accurately. Advanced VR systems also integrate spatial audio to enhance the immersive experience, replicating sounds such as birdsong, rustling leaves, and flowing water. These elements are crucial in providing a multisensory experience that can trigger the same psychological responses as actual natural environments (Mattila et al., 2020).

To ensure the environments are perceived as natural and restorative, designers must pay attention to details such as the natural progression of time (e.g., changing light conditions) and weather effects. The inclusion of interactive elements, where users can move around and interact with the environment, further enhances the sense of presence and engagement. Research shows that the more lifelike and interactive the VR environment, the more effective it is in promoting psychological restoration (de Kort et al., 2006).

5.3.3   *Psychological Benefits*

The psychological benefits of EPR using VR are significant and well-documented. One of the primary benefits is the reduction of stress and anxiety. VR environments that simulate natural settings have been shown to lower cortisol levels and induce a state of relaxation, similar to that experienced in real natural settings (Ulrich et al., 1991).

Additionally, VR-based EPR can enhance mood and overall emotional well-being. Exposure to restorative environments, even virtually, can reduce symptoms of depression and improve mood, providing a valuable tool for mental health interventions (Korpela et al., 2008). Cognitive benefits are also notable, as these environments help replenish attention and cognitive resources, leading to improved concentration and mental clarity (Berto, 2005).

VR offers the flexibility to tailor environments to individual needs, enhancing the effectiveness of the restoration. This customization can lead to higher engagement and motivation, crucial factors in both educational and therapeutic contexts. Regular exposure to these restorative VR environments can lead to sustained improvements in psychological well-being, making it a powerful tool for long-term mental health maintenance (Makransky & Lilleholt, 2018).

5.4   Case Study: Effects of an EPR Protocol with VR on Indicators of Demotivation in Children in México

5.4.1   Methodology

This study utilized a controlled experimental design with both an experimental group and a control group. The participants consisted of 30 boys and 30 girls, aged 8 to 12, selected through a public call on social media. The parents volunteered their children, and the inclusion criteria required that the children did not have any identified medical conditions or developmental disorders, had no prior exposure to virtual reality, and had not stayed up late before the study.

Instruments: Demotivation Measurement: The Generalized Demotivation Scale for Children and Adolescents (EDG-NA), validated for the Mexican population and Alertness Evaluation: RehaCom, validated for the Mexican population. Both scales were administered at the beginning (pre-test) and at the end (post-test) of the 21-day study period.

Procedure: After obtaining informed consent from the parents, each child completed the initial EDG-NA and RehaCom assessments. Those of the experimental group experienced a virtual reality (VR) session featuring Environmental Psychological Restoration (EPR) for 15 minutes. This VR experience, consisting of restorative landscapes from a validated image bank, was repeated daily for 21 days. On the final day, the same scales were administered again as a post-test to both groups.

5.4.2   Results

The study observed a significant increase in motivation within the experimental group that underwent the 21-day VR EPR experience. The children in the experimental group displayed higher levels of creativity, well-being, and eagerness to return to the laboratory each day. Significant differences were found between the experimental and control groups, indicating the effectiveness of the VR EPR protocol.

Demotivation (EDG-NA): The Generalized Demotivation Scale for Children and Adolescents (EDG-NA) scores range from 0 to 100, where higher scores indicate higher levels of demotivation.

RehaCom's alertness scale ranges from -3 to 3, with higher scores indicating better alertness.

Creativity and Well-Being (Observed)

Creativity: The experimental group showed significant improvement, with 85% of participants demonstrating higher creativity levels post-intervention. The control group showed no significant changes.

Well-Being: Observations indicated increased levels of well-being in the experimental group, with children displaying more smiles, enthusiasm, and eagerness to return to the laboratory each day. No significant changes were observed in the control group (Table 5.1).

5.4.3   Detailed Results

Reduction in Demotivation: the experimental group exhibited a significant reduction in demotivation scores from pre-test (Mean = 50.2, SD = 8.4) to post-test (Mean = 30.1, SD = 7.2), $t(29)$ = -9.12, $p < 0.001$. In contrast, the control group showed no significant change, $t(29)$ = -1.32, p = 0.20. This indicates that the VR-based EPR protocol effectively reduces demotivation in children, corroborating findings from previous studies that highlight the potential of immersive environments in enhancing motivation and engagement (Li et al., 2019).

Increase in Alertness: the alertness scores for the experimental group significantly increased from pre-test (Mean = -1.42, SD = 9.0) to post-test (Mean = 0.92, SD = 7.5), $t(29)$ = 8.45, $p < 0.001$. The control group, however, showed no significant change in alertness, $t(29)$ = 0.91, p = 0.37. This suggests that the protocol not only reduces demotivation but also enhances cognitive alertness, aligning with research indicating that VR can

Table 5.1   Results

Measure	Experimental Group Mean (SD)	Control Group Mean (SD)
Demotivation (EDG-NA) Pre-test	50.2 (8.4)	48.9 (7.9)
Demotivation (EDG-NA) Post-test	30.1 (7.2)	47.5 (8.1)
Alertness (RehaCom) Pre-test	-1.42 (9.0)	-1.26 (8.7)
Alertness (RehaCom) Post-test	0.92 (7.5)	-1.16 (8.4)
Creativity (Observed)	85% improved	No significant changes
Well-Being (Observed)	Increased smiles, enthusiasm	No significant changes

improve cognitive functions through engaging and stimulating environments (Rizzo et al., 2017).

Improvement in Creativity and Well-Being: qualitative observations showed that 85% of the children in the experimental group exhibited higher creativity post-intervention, and there were noticeable increases in well-being, such as more frequent smiling and expressions of joy. These findings, shown in Table 5.2, are consistent with literature suggesting that immersive VR environments can enhance creative thinking and emotional well-being by providing novel and stimulating experiences (Plante et al., 2003).

5.4.4   Conclusion

The Virtual Reality-based Environmental Psychological Restoration (EPR) protocol has shown significant efficacy in reducing demotivation and enhancing alertness in children. The experimental group demonstrated a notable decrease in demotivation scores and a substantial increase in alertness post-intervention. Qualitative observations also indicated improvements in creativity and well-being, suggesting that the VR-based EPR protocol not only addresses cognitive aspects but also positively influences emotional and creative dimensions.

5.4.5   Discussion

The results of this study highlight the transformative potential of VR-based EPR protocols in addressing demotivation and enhancing alertness among children. The significant reduction in demotivation scores and the increase in alertness suggest that VR can be an effective tool in educational and therapeutic settings. The observed improvements in creativity and

Table 5.2 Statistical analysis

Measure	Group	t value	p value
Demotivation (EDG-NA)	Experimental Group	–9.12	<0.001
Demotivation (EDG-NA)	Control Group	–1.32	0.2
Alertness (RehaCom)	Experimental Group	8.45	<0.001
Alertness (RehaCom)	Control Group	0.91	0.37

well-being further support the use of VR in holistic interventions that target cognitive, emotional, and creative development.

The control group's lack of significant change underscores the importance of the VR component in the intervention, suggesting that traditional methods may not be as effective in engaging children or altering cognitive and emotional states. These findings are in line with previous research that emphasizes the unique benefits of VR in providing immersive, engaging, and motivating environments (Anderson et al., 2017).

5.5   ADAPTABILITY AND PERSONALIZATION OF VR

Virtual Reality (VR) stands out for its remarkable ability to tailor experiences to individual needs and preferences, offering a highly personalized learning environment. This adaptability is particularly beneficial in educational settings, where each child's unique learning style, pace, and interests can significantly impact their educational outcomes.

5.5.1   *Personalization of Experiences*

5.5.1.1 *Customization of Learning Content*
One of the key advantages of VR is its ability to customize learning content to match the individual learning paths of students. For instance, VR applications can adjust the difficulty level of tasks based on real-time performance data, ensuring that students are continually challenged yet not overwhelmed. This dynamic adjustment promotes a more engaging and effective learning experience (Parsons & Cobb, 2011).

5.5.1.2 *Individual Preferences and Interests*
VR environments can be modified to reflect the personal interests of students, making learning more relatable and engaging. For example, a student interested in marine biology can explore underwater ecosystems in VR, while another student fascinated by astronomy can embark on virtual space missions. This alignment with personal interests not only enhances motivation but also deepens the learning experience by allowing students to explore topics they are passionate about (Merchant et al., 2014).

Flexible Learning Environments: VR can create flexible learning environments that cater to the sensory and cognitive needs of individual students. For instance, some students may benefit from a highly immersive and interactive environment, while others may need a more focused and distraction-free setting. VR can adjust the level of sensory input, interaction, and complexity to suit the learner's needs, providing an optimal learning experience for each student (Rizzo et al., 2004).

5.6   DIVERSITY OF LEARNING STYLES

VR's ability to cater to diverse learning styles is one of its most powerful features. Students have different ways of processing information and learning effectively. VR can address these differences in a variety of ways:

5.6.1   Visual Learners

VR provides rich, immersive visual experiences that can enhance understanding and retention for visual learners. For example, visual learners can benefit from 3D models, simulations, and interactive environments that illustrate complex concepts more vividly than traditional textbooks (Makransky & Lilleholt, 2018).

5.6.2   Auditory Learners

VR can integrate auditory elements such as narration, sound effects, and interactive dialogues to cater to auditory learners. This integration helps in reinforcing learning through hearing, complementing the visual and kinesthetic components of the VR experience (Clark & Mayer, 2016).

5.6.3   Kinesthetic Learners

VR is particularly effective for kinesthetic learners who learn best through hands-on experiences. VR simulations and interactive activities allow students to manipulate objects, conduct virtual experiments, and engage in physical activities within a virtual space, thereby enhancing their learning through movement and touch (Huang et al., 2019).

5.6.4   Impact on Motivation and Engagement

By addressing diverse learning styles, VR can significantly boost student motivation and engagement. When students interact with learning material in a way that aligns with their natural preferences, they are more likely to be engaged and motivated. This personalized approach can lead to better educational outcomes, as students are more inclined to participate actively and enjoy the learning process (Lamb et al., 2018).

5.7   Future Implications and Conclusions

5.7.1   Future of VR in Education and Therapy

5.7.1.1   Virtual Reality (VR) as a Fundamental Tool
As VR technology continues to evolve, its potential to transform education and therapy becomes increasingly apparent. VR's immersive capabilities allow for experiential learning and therapy that can address individual needs more effectively than traditional methods. For instance, in education, VR can simulate historical events, complex scientific phenomena, and far-reaching geographical explorations, making abstract concepts tangible and engaging for students (Huang et al., 2019). In therapy, VR can provide safe and controlled environments for exposure therapy, cognitive rehabilitation, and skill development for individuals with disabilities or mental health conditions (Rizzo et al., 2017).

5.7.1.2   Advancements in Technology
Future advancements in VR technology, such as improved hardware (e.g., lighter, more comfortable headsets) and more sophisticated software (e.g., AI-driven adaptive learning systems), will enhance the efficacy and accessibility of VR in educational and therapeutic settings. Innovations like haptic feedback, which allows users to "feel" virtual objects, and more natural user interfaces, such as gesture and voice controls, will further bridge the gap between virtual and real-world experiences (Anderson et al., 2017).

5.7.1.3   Scalability and Accessibility
As VR becomes more affordable, its integration into mainstream education and therapy will likely increase. Schools and therapy centers in underserved or remote areas could particularly benefit from VR, providing high-quality, interactive, and engaging learning and therapeutic experiences that might otherwise be unavailable (Merchant et al., 2014).

5.7.2   Recommendations

5.7.2.1   Implementation in Education

5.7.2.1.1   Teacher Training
Invest in comprehensive training programs for educators to effectively integrate VR into their teaching practices. This includes understanding the technology, developing VR lesson plans, and troubleshooting common issues (Clark & Mayer, 2016).

5.7.2.1.2   Curriculum Alignment
Ensure VR content aligns with curriculum standards and enhances existing educational goals. Use VR as a complementary tool to traditional teaching methods, rather than a replacement (Makransky & Lilleholt, 2018).

5.7.2.1.3   Student-Centered Design
Develop VR experiences that are adaptable to various learning styles and individual student needs. Include features that allow for personalization and flexibility to accommodate diverse educational contexts (Parsons & Cobb, 2011).

5.7.2.2   Implementation in Therapy

5.7.2.2.1   Therapist Training
Provide specialized training for therapists to use VR effectively in clinical settings. This should cover technical aspects, therapeutic techniques, and ethical considerations related to VR use (Rizzo et al., 2017).

5.7.2.2.2   Evidence-Based Practices
Integrate VR interventions that are supported by robust clinical evidence. Regularly update VR therapy protocols based on the latest research to ensure efficacy and safety (Li et al., 2019).

5.7.2.2.3   Patient-Centered Care
Design VR therapy programs that are tailored to the individual needs and preferences of patients. Incorporate feedback mechanisms to continually improve the therapeutic experience and outcomes (Anderson et al., 2017).

5.7.3   Conclusions

This chapter has explored the transformative potential of VR in education and therapy, highlighting its adaptability, personalization capabilities, and impact on motivation and engagement. Key points discussed include:

Personalization of Experiences: VR can be customized to match the individual needs and preferences of students and patients, making learning and therapy more engaging and effective (Merchant et al., 2014).

Diversity of Learning Styles: VR addresses diverse learning styles, enhancing educational outcomes by providing visual, auditory, and kinesthetic learning opportunities (Huang et al., 2019).

Future Implications: Advances in VR technology will likely increase its scalability and accessibility, making it a fundamental tool in both education and therapy (Rizzo et al., 2017).

5.8   Final Reflection

The potential of VR to revolutionize education and therapy is immense. Its ability to create immersive, engaging, and personalized experiences offers new possibilities for enhancing learning and therapeutic outcomes. As technology advances and becomes more accessible, the integration of VR into educational and therapeutic practices will likely become more widespread, offering benefits that were previously unimaginable. Embracing VR as a fundamental tool in these fields can lead to more effective and engaging ways to educate and heal, ultimately improving the quality of life for learners and patients alike.

References

Anderson, P. L., Price, M., Edwards, S. M., Obasaju, M. A., Schmertz, S. K., Zimand, E., & Calamaras, M. R. (2017). Virtual reality exposure therapy for social anxiety disorder: A randomized controlled trial. *Journal of Consulting and Clinical Psychology, 85*(3), 238–249. https://doi.org/10.1080/1040041 9.2017.1302741

Bandura, A. (1986). *Social foundations of thought and action: A social cognitive theory.* Prentice-Hall. https://doi.org/10.1016/0749-5978(86)90028-1

Berto, R. (2005). Exposure to restorative environments helps restore attentional capacity. *Journal of Environmental Psychology, 25*(3), 249–259. https://doi. org/10.1016/j.jenvp.2004.12.001

Clark, R. C., & Mayer, R. E. (2016). *E-learning and the science of instruction: Proven guidelines for consumers and designers of multimedia learning*. John Wiley & Sons. https://doi.org/10.1002/9781119239086

de Kort, Y. A. W., Meijnders, A. L., Sponselee, A. A. G., & Ijsselsteijn, W. A. (2006). What's wrong with virtual trees? Restoring from stress in a mediated environment. *Journal of Environmental Psychology, 26*(4), 309–320. https://doi.org/10.1016/j.jenvp.2005.12.003

Deci, E. L., Koestner, R., & Ryan, R. M. (1999). A meta-analytic review of experiments examining the effects of extrinsic rewards on intrinsic motivation. *Psychological Bulletin, 125*(6), 627–68. https://doi.org/10.1037/0033-2909.125.6.627

Dewey, J. (1938). *Experience and education*. Macmillan.

Dweck, C. S. (1986). Motivational processes affecting learning. *American Psychologist, 41*(10), 1040–1048. https://doi.org/10.1037/0022-3514.53.6.1024

Eccles, J. S., Adler, T. F., Futterman, R., Goff, S. B., Kaczala, C. M., Meece, J. L., & Midgley, C. (1983). Expectancies, values, and academic behaviors. *Achievement and Achievement Motives*, 75–146. https://doi.org/10.3102/00346543073001068

Fredricks, J. A., Blumenfeld, P. C., & Paris, A. H. (2004). School engagement: Potential of the concept, state of the evidence. *Review of Educational Research, 74*(1), 59–109. https://doi.org/10.3102/00346543074001059

Freina, L., & Ott, M. (2015). A literature review on immersive virtual reality in education: State of the art and perspectives. *Conference proceedings of eLearning and software for education (eLSE)*. https://doi.org/10.12753/2066-026X-15-020

Gaggioli, A., Chirico, A., Mazzoni, E., Milani, L., & Riva, G. (2015). A strategy for mental health and well-being: Positive technology and happy machines. *Frontiers in Psychology, 6*, 295. https://doi.org/10.3389/fpsyg.2015.00295

Huang, W. H. Y., Rauch, U., & Liaw, S. S. (2019). Investigating learners' attitudes toward virtual reality learning environments: Based on a constructivist approach. *Interactive Learning Environments, 28*(1), 1–14. https://doi.org/10.1109/TLT.2018.2868673

Kaplan, R., & Kaplan, S. (1989). *The experience of nature: A psychological perspective*. Cambridge University Press.

Kjellgren, A., & Buhrkall, H. (2010). A comparison of the restorative effect of a natural environment with that of a simulated natural environment. *Journal of Environmental Psychology, 30*(4), 464–472. https://doi.org/10.1016/j.jenvp.2010.01.011

Korpela, K. M., Ylén, M., Tyrväinen, L., & Silvennoinen, H. (2008). Determinants of restorative experiences in everyday favorite places. *Health & Place, 14*(4), 636–652. https://doi.org/10.1016/j.healthplace.2007.10.008

Lamb, R. L., Annetta, L. A., Firestone, J., & Etopio, E. (2018). A meta-analysis with examination of moderators of student cognition, affect, and learning outcomes while using serious educational games, serious games, and simulations. *Computers in Human Behavior, 80*, 158–167. https://doi.org/10.1007/s11423-017-9552-5

Lau, W. K., Chiu, T., Ho, K., Lo, S., & Luk, M. (2017). The effectiveness of virtual reality-based training on functional outcomes for children with cerebral palsy: A systematic review and meta-analysis. *Developmental Medicine & Child Neurology, 59*(7), 717–725. https://doi.org/10.1111/dmcn.13419

Li, B. J., Liu, Y., Jiang, X. L., Han, W., Zhang, J., & Zhao, Z. (2019). The impact of virtual reality games on the development of children's cognitive and emotional abilities: A longitudinal study. *Journal of Experimental Psychology: Human Perception and Performance, 45*(6), 778–790. https://doi.org/10.1037/xhp0000737

Makransky, G., & Lilleholt, L. (2018). A structural equation modeling investigation of the emotional value of immersive virtual reality in education. *Educational Technology Research and Development, 66*(5), 1141–1164. https://doi.org/10.1007/s11423-018-9579-6

Maskey, M., Lowry, J., Rodgers, J., McConachie, H., & Parr, J. R. (2014). Reducing anxiety in children with autism spectrum disorder using a virtual reality environment: A pilot study. *PLoS One, 9*(7), e100374. https://doi.org/10.1371/journal.pone.0100374

Mattila, O., Korhonen, A., Pöyry, E., Henttonen, P., Parvinen, P., & Mäkelä, T. (2020). Restoration in a virtual reality forest environment. *Frontiers in Psychology, 11*, 352. https://doi.org/10.3389/fpsyg.2020.00352

Merchant, Z., Goetz, E. T., Cifuentes, L., Keeney-Kennicutt, W., & Davis, T. J. (2014). Effectiveness of virtual reality-based instruction on students' learning outcomes in K-12 and higher education: A meta-analysis. *Computers & Education, 70*, 29–40. https://doi.org/10.1007/s11423-013-9327-9

Parsons, S., & Cobb, S. (2011). State-of-the-art of virtual reality technologies for children on the autism spectrum. *European Journal of Special Needs Education, 26*(3), 355–366. https://doi.org/10.1080/08856257.2011.593831

Plante, T. G., Aldridge, A., Su, D., Bogdan, R., Belo, M., & Kahn, K. (2003). Does virtual reality enhance the management of stress?: A meta-analytic review of the research. *Journal of Psychosomatic Research, 55*(3), 243–256. https://doi.org/10.1016/S0022-3999(03)00557-0

Riva, G., Bacchetta, M., Baruffi, M., Rinaldi, S., & Molinari, E. (2007). Virtual reality-based experiential cognitive treatment of obesity and binge-eating disorders. *Clinical Psychology & Psychotherapy: An International Journal of Theory & Practice, 10*(4), 241–253. https://doi.org/10.1002/cpp.328

Rizzo, A. A., Buckwalter, J. G., Neumann, U., Chua, C., Van Rooyen, A., Thiebaux, M., Humphrey, L., & Larson, P. (2004). Virtual environments for targeting cognitive processes: An overview of projects at the USC integrated media systems center. *Presence: Teleoperators and Virtual Environments, 13*(2), 150–169. https://doi.org/10.1145/1028014.1028016

Rizzo, A. S., Schultheis, M. T., Kerns, K. A., & Mateer, C. A. (2017). Analysis of the evidence: Clinical use of virtual reality technology in cognitive rehabilitation. *Frontiers in Psychology, 8,* 1223. https://doi.org/10.3389/fpsyg.2017.01223

Ryan, R. M., & Deci, E. L. (2000). Intrinsic and extrinsic motivations: Classic definitions and new directions. *Contemporary Educational Psychology, 25*(1), 54–67. https://doi.org/10.1006/ceps.1999.1020

Ulrich, R. S., Simons, R. F., Losito, B. D., Fiorito, E., Miles, M. A., & Zelson, M. (1991). Stress recovery during exposure to natural and urban environments. *Journal of Environmental Psychology, 11*(3), 201–230. https://doi.org/10.1016/S0272-4944(05)80184-7

CHAPTER 6

Processing Grief and Emotional Loss Through Mindfulness in Virtual Reality

Abstract This chapter examines the innovative integration of mindfulness and virtual reality (VR) to address grief and emotional loss. It begins by exploring the profound emotional challenges posed by grief and the urgent need for effective therapeutic methods. The chapter emphasizes the potential long-term mental health effects of unresolved grief, underscoring the necessity for novel interventions. Mindfulness, renowned for promoting emotional resilience and healing, is introduced as a crucial technique in this context. When combined with VR, mindfulness practices are transformed, offering immersive and introspective experiences for emotional processing. A detailed case study, "Effects of Mindfulness through Virtual Reality in Women with Recent Significant Emotional Losses," illustrates the efficacy of VR-based mindfulness interventions. This research demonstrates VR's unique ability to support individuals through the complexities of emotional loss by providing a flexible, personalized approach to mindfulness. The chapter highlights the adaptability and healing potential of VR, showcasing how it can address diverse emotional needs. By merging mindfulness with VR, a novel therapeutic avenue is presented, offering individuals a path toward emotional renewal and resilience. Ultimately, the chapter envisions a future where VR becomes a fundamental tool in emotional healing, providing solace and support within its immersive landscapes.

© The Author(s), under exclusive license to Springer Nature Switzerland AG 2024
D. M. Marchioro et al., *Virtual Reality: Unlocking Emotions and Cognitive Marvels*, Palgrave Studies in Cyberpsychology,
https://doi.org/10.1007/978-3-031-68196-7_6

Keywords Virtual reality • Mindfulness • Grief • Emotional loss • Emotional healing

6.1   INTRODUCTION

Grief and emotional loss are profound human experiences that affect individuals deeply, impacting their mental, emotional, and physical health. The process of grieving involves a complex interplay of emotions, including sadness, anger, guilt, and despair, which can lead to significant psychological distress (Bonanno & Kaltman, 2001). Emotional loss can result from various life events such as the death of a loved one, divorce, job loss, or other significant changes. It is essential to understand grief as a multifaceted process that requires comprehensive approaches to support individuals effectively (Stroebe et al., 2017).

Traditional therapeutic approaches to grief, including talk therapy, support groups, and medication, have been beneficial for many individuals. However, these methods often fall short of addressing the diverse needs and preferences of all grieving individuals. Traditional methods can sometimes feel impersonal or fail to engage individuals on a deeper emotional level, leading to insufficient emotional processing and prolonged grief (Shear, 2010). There is a growing recognition of the need for innovative therapeutic approaches that can provide more personalized, engaging, and effective support (Neimeyer, 2014).

Virtual Reality (VR) has emerged as a revolutionary tool in modern therapy, offering unique advantages over traditional methods. VR technology creates immersive and interactive environments that can enhance therapeutic experiences by providing a sense of presence and engagement that traditional methods often lack (Riva et al., 2019). In the context of grief and emotional loss, VR can simulate comforting and safe environments that facilitate emotional processing and healing. VR's ability to create controlled, customizable, and repeatable environments makes it an invaluable tool for therapists (Cipresso et al., 2018).

VR therapy can provide experiences that are otherwise difficult to achieve in real life, such as visiting a serene forest or a beach, which can be particularly beneficial for those unable to access such environments. This immersive aspect of VR can help individuals feel more connected to their emotions and facilitate a deeper level of mindfulness and emotional processing (Botella et al., 2017). As VR technology continues to advance and become more accessible, its potential applications in therapy expand, offering new possibilities for treating a wide range of psychological issues, including grief and emotional loss (Gonçalves et al., 2020).

This chapter will explore the integration of mindfulness and VR as a novel approach to processing grief and emotional loss. The chapter is structured as follows:

Exploring the Therapeutic Applications of VR for Grief Processing: This section examines traditional grief therapy methods, their limitations, and how VR offers unique benefits such as immersive healing environments, customization, presence, and emotional regulation. Real-world applications and success stories will also be discussed.

Utilizing VR for Mindfulness-Based Interventions: this section provides an introduction to mindfulness, its principles, and its benefits. It discusses how mindfulness can be applied in grief therapy and outlines best practices for designing mindful VR experiences, including calming visuals, ambient sounds, and guided meditation scripts. The structure and execution of VR mindfulness protocols, as well as technological considerations, will be detailed.

Case Study "Effects of Mindfulness through Virtual Reality in Women with Recent Significant Emotional Losses": This section presents a detailed case study that includes the study design and methodology, a description of the VR mindfulness program, participant experiences, emotional and psychological impact, and both quantitative and qualitative outcomes. The results will be analyzed and discussed, with implications for practice and suggestions for future research.

The chapter concludes with a summary of key points, the importance of VR in grief therapy, and the future potential of VR mindfulness interventions.

6.2   Exploring the Therapeutic Applications of VR for Grief Processing

6.2.1   Traditional Grief Therapy Methods: Limitations and Challenges

Traditional grief therapy methods, such as talk therapy and cognitive-behavioral therapy (CBT), have long been the cornerstone of addressing grief and emotional loss. These methods typically involve structured sessions with a therapist, where individuals discuss their feelings, thoughts, and experiences related to their loss. While effective for many, these methods present several limitations and challenges that can hinder their effectiveness for some individuals.

6.2.1.1   Reluctance to Verbalize Emotions

One significant limitation of traditional grief therapy is its heavy reliance on verbal communication. Many individuals find it difficult to articulate their deep-seated emotions, particularly when it comes to the complex and often overwhelming feelings associated with grief (Neimeyer, 2001). This reluctance can stem from cultural norms, personal discomfort, or simply a lack of words to express their experiences. As a result, therapy sessions may not fully capture the emotional depth and nuance of the individual's grief, leading to an incomplete or superficial processing of their loss.

6.2.1.2   Variability in Therapeutic Outcomes

The effectiveness of traditional grief therapy can vary widely based on numerous factors, including the individual's personality, their relationship with the therapist, and the specific techniques employed (Jordan & Neimeyer, 2003). For some, the structured, cognitive approach of CBT might not resonate, leaving them feeling disconnected or unsupported. Furthermore, the therapeutic alliance, or the bond between therapist and client, plays a crucial role in the success of therapy. A mismatch in this alliance can result in reduced engagement and lower therapy efficacy.

6.2.1.3   Time Constraints and Accessibility

Traditional grief therapy often requires regular, in-person sessions, which can be challenging for individuals with time constraints or those living in remote areas (Aoun et al., 2015). The need for consistent appointments can add to the stress, particularly if the individual is juggling multiple responsibilities or facing logistical barriers to accessing therapy. Additionally, the cost of ongoing therapy can be prohibitive for some, limiting access to those who may benefit the most.

6.2.1.4 Limited Engagement and Motivation

Maintaining motivation and engagement over the long term is another challenge of traditional grief therapy. Grieving individuals might find it difficult to stay committed to the therapeutic process, especially if they do not perceive immediate benefits. The passive nature of some traditional methods can also contribute to disengagement. Without active and stimulating components, therapy sessions may become monotonous, leading to decreased motivation and participation over time (Stroebe et al., 2007).

6.2.1.5 Emotional Safety and Retraumatization

Traditional therapy methods sometimes inadvertently risk retraumatizing individuals by encouraging them to repeatedly revisit painful memories and experiences (Pearlman & Saakvitne, 1995). While discussing one's loss is a critical part of the healing process, there is a fine balance between beneficial exposure and harmful retraumatization. Therapists must be highly skilled in managing this balance, but not all practitioners have the necessary training or experience, which can compromise the safety and effectiveness of the therapy.

6.2.1.6 Insufficient Tools for Emotional Regulation

Traditional grief therapy can sometimes fall short of providing immediate tools for emotional regulation. While cognitive strategies and talking can be helpful, they may not always offer the immediate relief needed during intense moments of grief. This gap can leave individuals feeling overwhelmed and unsupported in managing their emotions outside of therapy sessions (Gross, 2002).

6.2.2 VR as a Therapeutic Tool: Advantages and Unique Benefits

Virtual Reality (VR) offers several advantages over traditional grief therapy methods. VR creates immersive healing environments that can engage users more deeply than traditional methods. One significant benefit is the ability to simulate different environments and scenarios that can evoke emotional responses and facilitate emotional processing (Riva et al., 2016a).

This immersion can make therapeutic interventions more impactful and memorable.

Moreover, VR allows for high levels of customization and personalization. Therapists can tailor VR experiences to meet the specific needs of each individual, enhancing the relevance and effectiveness of the therapy (Maples-Keller et al., 2017). For example, a VR environment can be designed to simulate a place of personal significance, which can help individuals process their grief in a more meaningful context.

6.2.2.1 Immersive Healing Environments

Virtual Reality (VR) creates immersive healing environments that significantly enhance the therapeutic experience. Unlike traditional therapy, which relies on verbal communication and abstract visualization, VR immerses individuals in simulated environments that can evoke strong emotional responses and facilitate deeper emotional processing (Riva et al., 2016). These immersive experiences can make therapeutic interventions more impactful and memorable. For instance, VR can simulate serene natural settings such as a beach or forest, providing a sense of calm and relaxation that can help individuals open up more during therapy sessions, thereby enhancing the therapeutic process.

6.2.2.2 Customization and Personalization

One of the standout features of VR therapy is its high level of customization and personalization. Therapists can design VR experiences tailored to the specific needs of each individual, enhancing the relevance and effectiveness of the therapy (Maples-Keller et al., 2017). For example, VR can simulate environments or scenarios of personal significance to the individual, such as a beloved family home or a favorite vacation spot. This customization can make therapy more meaningful and impactful by allowing individuals to confront and process their grief in familiar and significant contexts. The ability to personalize VR environments ensures that therapy is not a one-size-fits-all approach but rather a unique experience suited to each person's emotional and psychological needs.

6.2.2.3 The Role of Presence

Presence, or the sense of being physically present in a particular environment, is a critical element of VR therapy. This feeling of presence can make therapeutic interventions more engaging and effective by helping individuals feel more connected to the virtual environment (Slater & Wilbur,

1997). In VR, the immersive experience can replicate the sensation of actually being in a place where individuals can safely explore their emotions. This sense of presence can enhance the emotional impact of the therapy, making it easier for individuals to access and process their feelings. The heightened engagement and emotional connection provided by presence in VR can lead to more profound therapeutic outcomes compared to traditional methods.

6.2.2.4 Emotional Regulation

VR therapy also offers significant benefits for emotional regulation. By providing a controlled and safe environment, VR allows individuals to confront and manage their emotions without the immediate pressures and distractions of the real world (Gorini & Riva, 2008). For example, VR environments can include calming elements such as soothing visuals and ambient sounds that help individuals regulate their emotions during therapy sessions. This controlled exposure can be particularly beneficial for individuals experiencing intense grief, as it provides a safe space to process their emotions without becoming overwhelmed. The ability to practice emotional regulation in a VR environment can translate to better emotional control in real-life situations.

6.2.2.5 Flexibility and Accessibility

VR therapy also offers increased flexibility and accessibility compared to traditional therapy methods. VR can be used in various settings, including at home, making therapy more accessible to individuals who may have difficulty attending regular in-person sessions due to geographical or time constraints (Freeman et al., 2017). This flexibility can be particularly beneficial for individuals in remote areas or those with limited mobility, as it reduces barriers to accessing effective therapeutic interventions. Additionally, VR can be a cost-effective alternative to traditional therapy, providing high-quality therapeutic experiences without the need for extensive travel or expensive in-person sessions.

6.2.2.6 Enhanced Engagement and Motivation

The interactive and immersive nature of VR can significantly enhance engagement and motivation in therapy. Traditional therapy can sometimes feel passive or monotonous, leading to disengagement over time. In contrast, VR therapy provides a dynamic and stimulating environment that can keep individuals more engaged and motivated throughout their therapeutic journey (Maples-Keller et al., 2017). This increased engagement can lead to more consistent and sustained participation in therapy, improving overall outcomes. The novelty and interactivity of VR can make therapy sessions more enjoyable and interesting, encouraging individuals to commit to and actively participate in their treatment.

6.2.3   Examples and Case Studies: Real-World Applications and Success Stories

The integration of VR into grief therapy offers numerous benefits, including enhanced engagement, customization, and emotional regulation. These advantages make VR a promising tool for addressing the complex and multifaceted nature of grief. By leveraging the immersive and adaptable nature of VR, therapists can provide more effective and personalized interventions, ultimately helping individuals navigate their emotional journeys with greater support and resilience (Table 6.1).

6.3   Utilizing VR for Mindfulness-Based Interventions

VR offers a powerful tool for delivering mindfulness-based interventions, especially in the context of grief therapy. By providing an immersive and controlled environment, VR can enhance the effectiveness of mindfulness practices, helping individuals process their grief and develop healthier coping strategies.

6.3.1   Introduction to Mindfulness: Principles and Benefits

Mindfulness is a mental practice that involves focusing one's attention on the present moment while calmly acknowledging and accepting one's feelings, thoughts, and bodily sensations.

6.3.1.1   Definition and Historical Background

Mindfulness is a mental practice that involves paying deliberate attention to the present moment without judgment. Rooted in ancient Buddhist traditions, mindfulness has been practiced for over 2500 years. It was originally intended as a path to enlightenment and spiritual awakening. In recent decades, mindfulness has been integrated into modern psychology

Table 6.1 Examples and cases

Case Study	Description	Key Findings	Implications for Grief Therapy
Repetto et al. (2013)	This study focused on using VR to help individuals with generalized anxiety disorder relive and reframe significant memories. Participants were exposed to VR scenarios that recreated personal memories, allowing gradual and therapeutic emotional processing	Participants experienced significant reductions in symptoms of anxiety and depression. The controlled exposure to VR facilitated emotional closure and effective emotional regulation	Highlights VR's potential to help individuals process grief by confronting and reframing painful memories in a safe, controlled manner, facilitating emotional healing and closure
Riva et al. (2016)	Explored the impact of VR-based therapy on individuals dealing with significant emotional loss. Participants engaged in VR sessions featuring calming environments and personalized therapeutic exercises	Participants reported significant improvements in emotional well-being, with notable reductions in symptoms of depression and anxiety. The immersive VR experience enhanced engagement and emotional processing	Demonstrates VR's effectiveness in grief therapy, showing how immersive, personalized VR environments can improve emotional regulation and overall mental health
Gorini & Riva, (2008)	Investigated VR's role in anxiety disorders, providing controlled exposure to anxiety-provoking scenarios. Participants gradually confronted their fears in VR, leading to improved emotional regulation	Significant reduction in anxiety symptoms, increased emotional resilience, and improved coping strategies	Suggests VR's potential in grief therapy to help individuals gradually confront and process grief-related anxieties, building emotional resilience and coping skills

(*continued*)

Table 6.1 (continued)

Case Study	Description	Key Findings	Implications for Grief Therapy
Botella et al. (2010)	Focused on the use of VR in treating PTSD by allowing patients to relive traumatic events in a controlled environment. The study aimed to reduce the emotional impact of these memories through gradual exposure	Participants showed decreased PTSD symptoms and improved emotional stability. The controlled VR environment helped in safely processing traumatic memories	Highlights how VR can be adapted for grief therapy to safely revisit and process traumatic grief experiences, reducing emotional distress and improving mental stability
Freeman et al. (2017)	Explored VR's application in various mental health disorders, including its use for therapeutic engagement and emotional processing. The study emphasized VR's flexibility and accessibility	Enhanced engagement and motivation in therapy, improved emotional regulation, and better therapeutic outcomes. VR's flexibility allowed for effective therapy across diverse settings	Underlines VR's adaptability and accessibility in grief therapy, making it a viable option for diverse populations, including those with limited access to traditional therapy
Maples-Keller et al. (2017)	Examined the use of VR technology in treating anxiety and other psychiatric disorders. The study highlighted VR's ability to create immersive and engaging therapeutic environments	VR therapy led to significant reductions in anxiety and improvements in emotional well-being. The immersive nature of VR enhanced therapeutic engagement and outcomes	Suggests VR's utility in grief therapy by providing engaging and immersive environments that enhance emotional processing and therapeutic outcomes
Slater et al. (2006)	Investigated the use of VR for public speaking anxiety, using VR to simulate public speaking scenarios. The study aimed to improve participants' confidence and reduce anxiety through repeated exposure	Participants showed increased confidence and reduced anxiety related to public speaking. The realistic VR scenarios provided effective exposure therapy	Demonstrates VR's potential for grief therapy by providing realistic simulations that help individuals confront and process grief-related fears and anxieties

(continued)

Table 6.1 (continued)

Case Study	Description	Key Findings	Implications for Grief Therapy
Garcia-Palacios et al. (2007)	Explored the use of VR exposure therapy for individuals with PTSD. The study used VR to simulate traumatic events in a controlled and therapeutic manner	Significant reductions in PTSD symptoms and improvements in emotional regulation. VR provided a safe space for individuals to process traumatic memories	Highlights VR's application in grief therapy by offering a safe and controlled environment for processing traumatic grief experiences, reducing symptoms, and improving emotional well-being

and medicine, primarily through the work of Jon Kabat-Zinn, who developed the Mindfulness-Based Stress Reduction (MBSR) program in the late 1970s. This program aimed to treat patients with chronic pain and stress-related conditions by using mindfulness techniques to enhance their quality of life (Kabat-Zinn, 1990).

6.3.1.2   Core Principles of Mindfulness

The practice of mindfulness is guided by several core principles, which are fundamental to its application and benefits:

Non-judgment: Observing experiences without labeling them as good or bad. This principle encourages acceptance and reduces the tendency to react negatively to unpleasant experiences.

Patience: Understanding that thoughts and emotions will change over time and allowing experiences to unfold in their own time.

Beginner's Mind: Viewing each moment as fresh and unique, without preconceived notions. This helps in experiencing life with curiosity and openness.

Trust: Developing a sense of confidence in oneself and one's feelings. Trusting one's inner wisdom and experiences.

Non-striving: Focusing on the present moment without trying to achieve any specific outcome or goal. This principle emphasizes being rather than doing.

Acceptance: Acknowledging and accepting things as they are in the present moment. Acceptance is crucial for dealing with situations that are beyond our control.

Letting Go: Releasing attachment to thoughts, emotions, and experiences. This principle involves not clinging to what is pleasant or pushing away what is unpleasant.

6.3.1.3 Benefits of Mindfulness

Mindfulness practice offers a wide range of psychological, emotional, and physical benefits. Extensive research has demonstrated its effectiveness in improving mental health and overall well-being.

Stress Reduction: Mindfulness helps reduce stress by promoting relaxation and reducing the physiological impact of stress on the body. It lowers cortisol levels and helps in managing the body's stress response (Chiesa & Serretti, 2009).

Improved Emotional Regulation: Practicing mindfulness enhances the ability to manage and respond to emotions in a balanced way. It helps individuals recognize their emotional patterns and respond rather than react impulsively (Roemer et al., 2009).

Enhanced Cognitive Function: Mindfulness training has been shown to improve attention, concentration, and executive functioning. It enhances cognitive flexibility and working memory, making it easier to focus and make decisions (Zeidan et al., 2010).

Reduction in Symptoms of Anxiety and Depression: Mindfulness-based interventions have been effective in reducing symptoms of anxiety and depression. By fostering a non-judgmental awareness of thoughts and feelings, mindfulness helps break the cycle of rumination and worry (Hofmann et al., 2010).

Increased Self-Awareness: Mindfulness promotes a deeper understanding of oneself by encouraging introspection and self-reflection. It helps individuals become more aware of their thoughts, feelings, and behaviors, leading to greater self-acceptance and personal growth (Brown & Ryan, 2003).

Better Physical Health: Regular mindfulness practice has been linked to various physical health benefits, including reduced blood pressure, improved immune function, and better sleep quality. It also helps in managing chronic pain and other medical conditions (Goyal et al., 2014).

Enhanced Interpersonal Relationships: Mindfulness fosters empathy, compassion, and non-reactive communication, which are essential for healthy relationships. It improves emotional intelligence and helps individuals respond more skillfully to interpersonal conflicts (Carson et al., 2007).

6.3.1.4 Mechanisms of Mindfulness

The effectiveness of mindfulness can be attributed to several underlying mechanisms:

Attention Regulation: Mindfulness training enhances the ability to sustain attention and shift focus when necessary. This improved attentional control is crucial for managing stress and emotional responses (Jha et al., 2007).

Body Awareness: By focusing on bodily sensations, mindfulness increases body awareness. This heightened awareness helps individuals recognize physical manifestations of stress and emotions, allowing for early intervention and better self-care (Mehling et al., 2011).

Emotional Regulation: Mindfulness promotes emotional regulation by increasing awareness of emotional triggers and reducing automatic reactivity. It helps individuals observe their emotions without getting overwhelmed, leading to more adaptive responses (Goldin & Gross, 2010).

Change in Perspective on the Self: Mindfulness encourages a shift in perspective from a fixed, self-centered viewpoint to a more flexible and interconnected understanding of the self. This change promotes self-compassion and reduces self-criticism (Vago & Silbersweig, 2012).

6.3.1.5 Applications of Mindfulness in Various Fields

Mindfulness has been integrated into various fields and practices due to its wide-ranging benefits:

Clinical Psychology and Psychiatry: Mindfulness-based interventions, such as MBSR and Mindfulness-Based Cognitive Therapy (MBCT), are widely used to treat conditions like depression, anxiety, PTSD, and substance abuse (Segal et al., 2012).

Education: Mindfulness programs in schools help students improve focus, emotional regulation, and academic performance. They also reduce stress and promote a positive learning environment (Zenner et al., 2014).

Workplace: Mindfulness training in the workplace enhances employee well-being, reduces burnout, and improves productivity. It also fosters better communication and teamwork (Hyland et al., 2015).

Healthcare: Healthcare professionals use mindfulness to manage stress, prevent burnout, and improve patient care. It enhances their ability to remain present and compassionate with patients (Irving et al., 2009).

Sports and Performance: Athletes and performers use mindfulness to improve concentration, reduce performance anxiety, and enhance overall performance. It helps them stay focused and calm under pressure (Gardner & Moore, 2012).

6.3.2   Mindfulness and Grief: Application in Grief Therapy

Grief is a complex and multifaceted emotional response to loss, often accompanied by intense feelings of sadness, anger, and confusion. Traditional grief therapy methods can be significantly enhanced by incorporating mindfulness practices, which help individuals process their emotions more effectively and find peace amidst their grief. Mindfulness-based interventions teach individuals to observe their emotions without judgment, allowing them to experience their grief fully without becoming overwhelmed (Chambers et al., 2009).

In the context of grief therapy, mindfulness practices can help individuals accept their loss and integrate it into their lives. These practices improve emotional regulation, reduce physical symptoms of grief (such as insomnia and fatigue), and develop coping strategies (Garland et al., 2011).

6.3.3   Designing Mindful VR Experiences: Best Practices and Considerations

VR offers a unique and powerful platform for delivering mindfulness-based interventions. The immersive nature of VR can enhance the effectiveness of mindfulness practices by providing a controlled and engaging environment. Here are some best practices for designing mindful VR experiences:

Calming Visuals: the visual component of VR is crucial for creating a calming and relaxing environment. Natural landscapes, such as beaches, forests, and mountains, can be particularly effective. These visuals should be immersive and realistic, helping individuals feel as though they are truly present in a peaceful setting. Studies have shown that exposure to nature in VR can significantly reduce stress and improve mood (Browning et al., 2020).

Ambient Sounds: sound is a powerful tool for enhancing mindfulness practices. Incorporating ambient sounds, such as flowing water, birdsong, and gentle breezes, can help individuals relax and focus their attention on the present moment. These sounds should be continuous and soothing, avoiding any abrupt or jarring noises that could disrupt the mindfulness practice (Anderson et al., 2017).

Guided Meditation Scripts: guided meditations can be highly effective in helping individuals navigate their mindfulness practice. These scripts should be carefully crafted to include clear and calming instructions, guiding individuals through mindfulness exercises such as breathing techniques, body scans, and visualization exercises. Research indicates that guided meditation in VR can enhance the overall experience and efficacy of mindfulness practices (Tarrant et al., 2018).

6.3.4 Structure and Execution of VR Mindfulness Protocols

Implementing VR mindfulness interventions requires a well-structured protocol to ensure maximum effectiveness:

Session Preparation: prior to the session, ensure that the VR equipment is functioning correctly and the environment is set up to minimize distractions. The participant should be briefed on what to expect during the session.

Session Structure: a typical session might begin with a brief introduction to mindfulness and the specific exercises to be conducted. This can be followed by the immersive VR experience, incorporating calming visuals, ambient sounds, and guided meditation scripts. Sessions should be designed to last between 20–30 minutes to maintain engagement and effectiveness.

Follow-Up: after the VR session, participants should be given time to reflect on their experience. This can include discussing their feelings, any insights gained, and how they might apply mindfulness techniques in their daily lives. Follow-up sessions can help reinforce the benefits and provide ongoing support.

6.3.5 Technological Considerations

Implementing VR mindfulness interventions involves several technological considerations

Hardware: ensure the VR hardware is of high quality to provide an immersive experience. This includes VR headsets, motion controllers, and potentially haptic feedback devices to enhance the sensory experience.

Software: the VR content should be specifically designed for mindfulness interventions, with high-quality visuals and sound. Customizable features allow for personalization to meet individual needs.

Accessibility: make sure the VR setup is accessible to all users, including those with physical disabilities. This may involve adjustable settings and ensuring that the software is user-friendly.

6.4   Case Study "Effects of Mindfulness through Virtual Reality in Women with Recent Significant Emotional Losses"

This case study examines the effects of mindfulness practices delivered through virtual reality (VR) technology on women who have recently experienced significant emotional losses. The study aims to explore the emotional and psychological impacts of VR mindfulness programs, focusing on both quantitative and qualitative outcomes. By integrating immersive technology with therapeutic mindfulness practices, this study provides insights into innovative approaches to mental health support.

6.4.1   Study Design and Methodology

The study employed a mixed-methods approach to gather both quantitative and qualitative data. The participants included 20 women aged between 25 and 50, who had experienced significant emotional losses within the last six months. The recruitment was conducted through local support groups and online advertisements. Participants were randomly assigned to either the intervention group (receiving the VR mindfulness program) or the control group (receiving traditional mindfulness sessions without VR). The intervention lasted for eight weeks, with participants engaging in three 30-minute sessions per week. Pre- and post-intervention assessments were conducted to measure emotional and psychological impacts using validated scales such as the Beck Depression Inventory (BDI), the State-Trait Anxiety Inventory (STAI), and the Perceived Stress Scale (PSS).

6.4.2 Description of the VR Mindfulness Program

The VR mindfulness program was developed to create a highly immersive and interactive environment, designed to enhance the traditional mindfulness experience. The program included various mindfulness exercises such as guided meditations, body scans, and mindful breathing, all set in serene virtual environments like forests, beaches, and mountains.

Participants wore VR headsets and were guided through the exercises by an instructor. The use of 360-degree video and spatial audio aimed to create a sense of presence and tranquility, helping participants to fully engage with the mindfulness practices.

6.4.3 Participant Experiences

Qualitative data were gathered through semi-structured interviews conducted at the end of the intervention. Participants in the VR group reported feeling more immersed and engaged compared to traditional mindfulness practices. Many noted the VR environment helped them to detach from daily stressors and focus on the present moment. The qualitative data collected from the semi-structured interviews revealed rich insights into the participants' experiences with the VR mindfulness program.

6.4.4 Quantitative and Qualitative Outcomes

Quantitative outcomes are reported in Table 6.2.

6.4.4.1 Qualitative Outcomes

Participants in the VR group reported higher levels of engagement and satisfaction with the mindfulness practice. They also expressed that the immersive experience helped them to better cope with their emotional losses.

Immersive Engagement: participants consistently reported that the VR environment facilitated a deeper level of engagement with mindfulness practices. The immersive nature of VR helped them to fully concentrate on the exercises without being easily distracted. One participant shared, "When I put on the VR headset, it felt like I was transported to another world. It was much easier to focus on the meditation compared to when I tried it on my own at home."

Table 6.2 Quantitative Outcomes

Measure	VR Group Pre-Intervention	VR Group Post-Intervention	Mean Change in VR Group	Control Group Pre-Intervention	Control Group Post-Intervention	Mean Change in Control Group	Statistical Significance (p value)
Beck Depression Inventory (BDI)	28	16	-12 points	27	22	-5 points	<0.05
State-Trait Anxiety Inventory (STAI)	55	40	-15 points	54	46	-8 points	<0.05
Perceived Stress Scale (PSS)	30	20	-10 points	31	27	-4 points	<0.05

Enhanced Emotional Connection: the serene virtual settings, such as peaceful beaches and tranquil forests, created a strong emotional connection for many participants. These environments helped evoke feelings of calmness and safety, which were crucial for those dealing with recent emotional losses. A participant noted, "The VR beach was so calming; I could hear the waves and see the sunset. It felt like a real escape from my daily struggles."

Reduced Anxiety and Stress: many participants highlighted a significant reduction in anxiety and stress levels during and after the VR sessions. The combination of guided mindfulness exercises and immersive environments appeared to have a soothing effect on their emotional state. One woman mentioned, "After each session, I felt a noticeable decrease in my anxiety. The VR experience allowed me to let go of my worries and just be present."

Sense of Presence and Reality: the realism of the VR experiences contributed to a strong sense of presence, which made the mindfulness practices more impactful. Participants felt as if they were truly in the virtual environments, which enhanced the effectiveness of the mindfulness exercises. "It was incredible how real it felt. I could almost feel the breeze and the warmth of the sun in the virtual forest," said one participant.

Coping Mechanism: several participants described the VR mindfulness program as an effective coping mechanism for their grief and emotional pain. The virtual environments provided a space where they could process their emotions without the distractions of their everyday lives. One participant explained, "The VR sessions became a safe haven for me. They helped me to process my grief in a way that felt manageable and supportive."

Positive Feedback on Program Structure: the structured nature of the VR mindfulness program, with regular sessions and guided exercises, was appreciated by the participants. They felt that consistency and guidance were crucial in helping them stick with the practice. "Having a set schedule and a virtual guide made a big difference. It was easier to commit to the practice knowing what to expect," commented one participant.

6.4.5   *Discussion and Implications for Practice*

The findings from this study indicate that VR mindfulness programs can be a powerful tool in reducing symptoms of depression, anxiety, and perceived stress in women who have recently experienced significant emotional losses. The immersive nature of VR technology appears to significantly enhance the therapeutic effects of mindfulness practices, providing a more engaging and effective intervention compared to traditional methods.

The realistic virtual environments created by VR technology enable participants to engage more deeply with mindfulness exercises. This heightened engagement likely contributes to the greater reductions in depressive symptoms, anxiety, and stress observed in the VR group. The sense of presence and immersion in the VR settings allows participants to focus more effectively on mindfulness practices, leading to better emotional and psychological outcomes (Bailenson, 2018).

The serene virtual environments used in the VR mindfulness program provide a significant source of emotional relief and comfort for participants. By creating a sense of escape from their everyday stressors and emotional pain, VR helps participants to relax and engage more fully in the mindfulness exercises. This emotional relief is critical for individuals dealing with grief and significant emotional losses (Gorini & Riva, 2008).

For practitioners, incorporating VR mindfulness programs into therapeutic practices offers a novel approach to mental health support. The structured format and guided sessions of the VR program ensure consistency and adherence, which can be challenging with traditional mindfulness practices. Additionally, the use of VR technology can make mindfulness practices more accessible and appealing to a broader range of individuals, potentially increasing the uptake and effectiveness of these interventions (Serino et al., 2014).

Future studies should consider larger sample sizes and diverse populations to validate these findings. Longitudinal studies could provide insights into the sustained effects of VR mindfulness programs. Additionally, exploring the integration of other therapeutic modalities with VR technology could further enhance mental health interventions.

6.4.6   *Suggestions for Future Research*

Future research should explore the long-term effects of VR mindfulness interventions. Longitudinal studies could provide insights into the sustainability of the benefits observed in the short term. Understanding how long the effects of VR mindfulness last can help in designing follow-up interventions or maintenance programs to ensure enduring benefits (Gutiérrez-Maldonado et al., 2015).

The current study focused on women who had experienced recent significant emotional losses. Future research should examine the efficacy of VR mindfulness programs in diverse populations, including men, individuals from different cultural backgrounds, and those with varying mental health conditions. This broader investigation can help in tailoring VR mindfulness programs to meet the specific needs of different groups (Maples-Keller et al., 2017).

Investigating the potential for integrating VR mindfulness with other therapeutic modalities, such as cognitive-behavioral therapy (CBT) or exposure therapy, could further enhance the effectiveness of mental health interventions. Combining VR mindfulness with other evidence-based therapies might provide a comprehensive approach to treating complex mental health issues (Freeman et al., 2017).

As VR technology continues to evolve, future research should explore the impact of advanced VR features, such as haptic feedback, biofeedback integration, and adaptive virtual environments, on the effectiveness of mindfulness interventions. These technological advancements could further enhance the immersive experience and therapeutic outcomes (Riva et al., 2016).

REFERENCES

Anderson, A. P., Mayer, M. D., Fellows, A. M., Cowan, D. R., Hegel, M. T., & Buckey, J. C. (2017). Relaxation with immersive natural scenes presented using virtual reality. *Aerospace Medicine and Human Performance, 88*(6), 520–526. https://doi.org/10.3357/AMHP.4743.2017

Aoun, S. M., Breen, L. J., Howting, D. A., Oliver, D., Henderson, R. D., & Edis, R. (2015). Who needs bereavement support? A population-based survey of bereavement risk and support need. *PLoS One, 10*(3), e0121101. https://doi.org/10.1371/journal.pone.0121101

Bailenson, J. (2018). Protecting nonverbal data tracked in virtual reality. *Nature Human Behaviour, 2*(7), 431–432. https://doi.org/10.1038/s41562-018-0400-5

Bonanno, G. A., & Kaltman, S. (2001). The varieties of grief experience. *Clinical Psychology Review, 21*(5), 705–734. https://doi.org/10.1016/S0272-7358(00)00062-3

Botella, C., García-Palacios, A., Baños, R. M., & Quero, S. (2010). Cybertherapy: Advantages, limitations, and ethical issues. *PsychNology Journal, 8*(1), 77–100. http://www.psychology.org/File/PNJ8(1)/PSYCHOLOGY_JOURNAL_8_1_BOTELLA.pdf

Botella, C., Serrano, B., Baños, R. M., & García-Palacios, A. (2017). Virtual reality exposure-based therapy for the treatment of post-traumatic stress disorder: A review of its efficacy, the adequacy of the treatment protocol, and its acceptability. *Neuropsychiatric Disease and Treatment, 13*, 2533–2545. https://doi.org/10.2147/NDT.S118620

Brown, K. W., & Ryan, R. M. (2003). The benefits of being present: mindfulness and its role in psychological well-being. *Journal of personality and social psychology, 84*(4), 822–848. https://doi.org/10.1037/0022-3514.84.4.822

Browning, M. H. E. M., Saeidi-Rizi, F., McAnirlin, O., Yoon, H., & Pei, Y. (2020). The role of methodological choices in the effects of experimental exposure to simulated natural landscapes on human health and cognitive performance: A systematic review. *Environment and Behavior, 52*(2), 111–143. https://doi.org/10.1177/0013916518800796

Carson, J. W., Carson, K. M., Gil, K. M., & Baucom, D. H. (2007). Self-expansion as a mediator of relationship improvements in a mindfulness intervention. *Journal of marital and family therapy, 33*(4), 517–528. https://doi.org/10.1111/j.1752-0606.2007.00035.x

Chambers, R., Gullone, E., & Allen, N. B. (2009). Mindful emotion regulation: An integrative review. *Clinical Psychology Review, 29*(6), 560–572. https://doi.org/10.1016/j.cpr.2009.06.005

Chiesa, A., & Serretti, A. (2009). Mindfulness-based stress reduction for stress management in healthy people: a review and meta-analysis. *Journal of alternative and complementary medicine, 15*(5), 593–600. https://doi.org/10.1089/acm.2008.0495

Cipresso, P., Giglioli, I. A. C., Raya, M. A., & Riva, G. (2018). The past, present, and future of virtual and augmented reality research: A network and cluster analysis of the literature. *Frontiers in Psychology, 9*, 2086. https://doi.org/10.3389/fpsyg.2018.02086

Freeman, D., Reeve, S., Robinson, A., Ehlers, A., Clark, D., Spanlang, B., & Slater, M. (2017). Virtual reality in the assessment, understanding, and treatment of mental health disorders. *Psychological Medicine, 47*(14), 2393–2400. https://doi.org/10.1016/S2215-0366(17)302998

Garcia-Palacios, A., Hoffman, H. G., Kwong See, S., Tsai, A., & Botella, C. (2007). Redefining therapeutic success with virtual reality exposure therapy. *Cyberpsychology & Behavior, 10*(3), 317–324. https://doi.org/10.1089/cpb.2006.9927

Gardner, F. L., & Moore, Z. E. (2012). Mindfulness and acceptance models in sport psychology: A decade of basic and applied scientific advancements. *Canadian Psychology/Psychologie Canadienne, 53*(4), 309–318. https://doi.org/10.1037/a0030220

Garland, E. L., Gaylord, S. A., & Fredrickson, B. L. (2011). Positive reappraisal mediates the stress-reductive effects of mindfulness: An upward spiral process. *Mindfulness, 2*(1), 59–67. https://doi.org/10.1007/s12671-011-0043-8

Goldin, P. R., & Gross, J. J. (2010). Effects of mindfulness-based stress reduction (MBSR) on emotion regulation in social anxiety disorder. *Emotion, 10*(1), 83–91. https://doi.org/10.1037/a0018441

Gonçalves, R., Pedrozo, A. L., Coutinho, E. S. F., Figueira, I., & Ventura, P. (2020). Efficacy of virtual reality exposure therapy in the treatment of PTSD: A systematic review. *PLoS One, 15*(12), e0243467. https://doi.org/10.1371/journal.pone.0243467

Gorini, A., & Riva, G. (2008). The potential of virtual reality as anxiety management tool: a randomized controlled study in a sample of patients affected by generalized anxiety disorder. *Trials, 9*(25). https://doi.org/10.1186/1745-6215-9-25.

Goyal, M., Singh, S., Sibinga, E. M., Gould, N. F., Rowland-Seymour, A., Sharma, R., Berger, Z., Sleicher, D., Maron, D. D., Shihab, H. M., Ranasinghe, P. D., Linn, S., Saha, S., Bass, E. B., & Haythornthwaite, J. A. (2014). Meditation programs for psychological stress and well-being: a systematic review and meta-analysis. *JAMA internal medicine, 174*(3), 357–368. https://doi.org/10.1001/jamainternmed.2013.13018

Gross, J. J. (2002). Emotion regulation: Affective, cognitive, and social consequences. *Psychophysiology, 39*(3), 281–291. https://doi.org/10.1017/S0048577201393198

Gutiérrez-Maldonado, J., Wiederhold, B. K., & Riva, G. (2015). Future directions: How virtual reality and Cyberpsychology can change the world. *Journal of Cybertherapy and Rehabilitation, 4*(4), 495–497. https://doi.org/10.1007/s10484-015-9272-2

Hofmann, S. G., Sawyer, A. T., Witt, A. A., & Oh, D. (2010). The effect of mindfulness-based therapy on anxiety and depression: A meta-analytic review. *Journal of Consulting and Clinical Psychology, 78*(2), 169–183. https://doi.org/10.1037/a0018555

Hyland, P., Lee, R., & Mills, M. (2015). Mindfulness at Work: A New Approach to Improving Individual and Organizational Performance. *Industrial and Organizational Psychology, 8*(4), 576–602. https://doi.org/10.1017/iop.2015.41

Irving, J. A., Dobkin, P. L., & Park, J. (2009). Cultivating mindfulness in health care professionals: a review of empirical studies of mindfulness-based stress reduction (MBSR). *Complementary Therapies in Clinical Practice, 15*(2), 61–6. https://doi.org/10.1016/j.ctcp.2009.01.002

Jha, A. P., Krompinger, J., & Baime, M. J. (2007). Mindfulness training modifies subsystems of attention. *Cognitive, affective & behavioral neuroscience, 7*(2), 109–119. https://doi.org/10.3758/cabn.7.2.109

Jordan, J. R., & Neimeyer, R. A. (2003). Does grief counseling work? *Death Studies, 27*(9), 765–786. https://doi.org/10.1080/07481180390229203

Kabat-Zinn, J. (1990). *Full catastrophe living: Using the wisdom of your body and mind to face stress, pain, and illness.* Delta.

Maples-Keller, J. L., Bunnell, B. E., Kim, S.-J., & Rothbaum, B. O. (2017). The use of virtual reality technology in the treatment of anxiety and other psychiatric disorders. *Clinical Psychology Review*, 56, 63–72. https://doi.org/10.1016/j.cpr.2017.02.004

Mehling, W. E., Wrubel, J., Daubenmier, J. J., Price, C. J., Kerr, C. E., Silow, T., Gopisetty, V., & Stewart, A. L. (2011). Body Awareness: a phenomenological inquiry into the common ground of mind-body therapies. *Philosophy, ethics, and humanities in medicine: PEHM*, 6, 6. https://doi.org/10.1186/1747-5341-6-6

Neimeyer, R. A. (2001). Meaning reconstruction & the experience of loss. *American Psychological Association.* https://doi.org/10.1037/10397-000

Neimeyer, R. A. (2014). The changing face of grief: Contemporary directions in theory, research, and practice. *Progress in Palliative Care*, 22(3), 125–130. https://doi.org/10.1179/1743291X13Y.0000000087

Pearlman, L. A., & Saakvitne, K. W. (1995). *Trauma and the therapist: Countertransference and vicarious traumatization in psychotherapy with incest survivors.* Norton & Company.

Repetto, C., Gaggioli, A., Pallavicini, F., Cipresso, P., & Riva, G. (2013). Virtual reality and mobile phones in the treatment of generalized anxiety disorders: A phase-2 clinical trial. *Personal and Ubiquitous Computing*, 17(2), 253–260. https://doi.org/10.1007/s00779-011-0467-0

Riva, G., Baños, R. M., Botella, C., Mantovani, F., & Gaggioli, A. (2016a). Transforming experience: The potential of augmented reality and virtual reality for enhancing personal and clinical change. *Frontiers in Psychiatry*, 7, 164. https://doi.org/10.3389/fpsyg.2016.01266

Riva, G., Baños, R. M., Botella, C., Wiederhold, B. K., & Gaggioli, A. (2016b). Virtual reality in the assessment and treatment of weight-related disorders. *Cyberpsychology, Behavior, and Social Networking*, 19(2), 69–73. https://doi.org/10.1089/cyber.2015.0267

Riva, G., Wiederhold, B. K., & Mantovani, F. (2019). Neuroscience of virtual reality: From virtual exposure to embodied medicine. *Cyberpsychology, Behavior, and Social Networking*, 22(1), 82–96. https://doi.org/10.1089/cyber.2017.29099.gri

Roemer, L., Lee, J. K., Salters-Pedneault, K., Erisman, S. M., Orsillo, S. M., & Mennin, D. S. (2009). Mindfulness and emotion regulation difficulties in generalized anxiety disorder: preliminary evidence for independent and overlapping contributions. *Behavior therapy*, 40(2), 142–154. https://doi.org/10.1016/j.beth.2008.04.001

Segal, Z., Williams, M., & Teasdale, J. (2012). *Mindfulness-Based Cognitive Therapy for Depression.* New York: Guilford Press.

Serino, S., Cipresso, P., Gaggioli, A., Pallavicini, F., & Riva, G. (2014). The potential of pervasive sensors and computing for positive technology: The interplay between human factors and technology. *Frontiers in Psychology, 5*, 1176. https://doi.org/10.3389/fpsyg.2014.01126

Shear, M. K. (2010). Complicated grief treatment: The theory, practice and outcomes. *Bereavement Care, 29*(3), 10–14. https://doi.org/10.1080/0268262 1.2010.522373

Slater, M., Pertaub, D. P., Barker, C., & Clark, D. M. (2006). An experimental study on fear of public speaking using a virtual environment. *Cyberpsychology & Behavior, 9*(5), 627–633. https://doi.org/10.1089/cpb.2006.9.627

Slater, M., & Wilbur, S. (1997). A framework for immersive virtual environments (FIVE): Speculations on the role of presence in virtual environments. *Presence: Teleoperators & Virtual Environments, 6*(6), 603–616. https://doi.org/10.1162/pres.1997.6.6.603

Stroebe, M., Schut, H., & Boerner, K. (2017). Cautioning health-care professionals: Bereaved persons are misguided through the stages of grief. *Omegai-Journal of Death and Dying, 74*(4), 455–473. https://doi.org/10.1177/0030222817691870

Stroebe, M., Schut, H., & Stroebe, W. (2007). Health outcomes of bereavement. *The Lancet, 370*(9603), 1960–1973. https://doi.org/10.1016/S0140-6736(07)61816-9

Tarrant, J., Viczko, J., & Cope, H. (2018). Virtual reality for anxiety reduction demonstrated by quantitative EEG: A pilot study. *Frontiers in Psychology, 9*, 1280. https://doi.org/10.3389/fpsyg.2018.01280

Vago, D. R., & Silbersweig, D. A. (2012). Self-awareness, self-regulation, and self-transcendence (S-ART): a framework for understanding the neurobiological mechanisms of mindfulness. *Frontiers in human neuroscience, 6*, 296. https://doi.org/10.3389/fnhum.2012.00296

Zeidan, F., Johnson, S. K., Diamond, B. J., David, Z., & Goolkasian, P. (2010). Mindfulness meditation improves cognition: evidence of brief mental training. *Consciousness and cognition, 19*(2), 597–605. https://doi.org/10.1016/j.concog.2010.03.014

Zenner. C., Herrnleben-Kurz, S., & Walach, H. (2014). Mindfulness-based interventions in schools-A systematic review and meta-analysis. *Frontiers in Psychology, 5*, 1–20. https://doi.org/10.3389/fpsyg.2014.00603

CHAPTER 7

Limitations and Future Developments

Abstract The discussion will consider technological constraints, accessibility issues, equipment costs, ecological validity, ethical challenges, privacy management, and the problem of technoreligions. By addressing these limitations, the aim is to develop a realistic and informed understanding of the current state of VR applications in psychology and to recognize that VR is more than just a set of technologies. Subsequently, possible future perspectives will be outlined, considering potential technological advancements, the development of a conscious approach to technologies, integration with artificial intelligence, new paradigms in research and therapy, and evolving ethical considerations. Citing Riva (*Comprehensive Clinical Psychology*, 91–105, 2022): The term "virtual reality" consists of two elements: "reality" (the actual state of things) and "virtual" (almost or almost as described). Thus, we can say that the concept of "virtual reality" essentially indicates a "quasi-real" or "similar to reality" reality, implying that VR represents a form of reality simulation. It is precisely this simulation capability that allows VR to pave the way for exciting application scenarios. The intention is to produce a complex reflection that starts from the limits and highlights, through critical analysis, the horizon toward which to aim for a conscious and politically correct use of new technologies.

169
D. M. Marchioro et al., *Virtual Reality: Unlocking Emotions and Cognitive Marvels*, Palgrave Studies in Cyberpsychology, https://doi.org/10.1007/978-3-031-68196-7_7

Keywords Technological constraints • Ethical challenges • Privacy management • Technological advancements • Future perspectives

7.1   Current Limitations of VR Applications in Psychology

While it is true that virtual reality (VR) technologies have opened new frontiers in psychological research and practice, they are not without limitations. By examining the current limitations, we can better understand the challenges that must be addressed to fully exploit the potential of VR. To date, there is still a need to develop awareness of this significant emerging topic. This section aims to delve into the current limitations while also providing insights into strategies to mitigate them and promote a healthy and balanced use of technology. By systematically analyzing the limitations, we can identify several important thematic areas to address:

High Costs: one of the main limitations is the costs associated with the purchase and subsequent maintenance of VR equipment. To offer immersive experiences, the devices used must be of high quality and constantly updated in both hardware and software. Equipment costs can be prohibitive for many academic and clinical institutions, and the problem is exacerbated for medium and small-sized institutions. In addition to the costs related to materials (hardware and/or software), the technical complexity of installing and managing VR systems requires highly qualified and trained personnel, adding further elements of cost and management. Moreover, the "almost exponential" speed at which technology evolves can render equipment obsolete, further increasing long-term costs. This rapid growth/evolution circuit intertwines both hardware and software development components.

Practical and Technological Limits: another significant obstacle, as mentioned earlier, is the complexity of the technology. VR requires specific technical skills to be used efficiently and effectively, meaning that psychologists and other mental health professionals must receive adequate training. The training process can be costly and time-consuming, reducing the ability to quickly adopt new technologies and their relative advancements. Furthermore, the lack of standardization in VR protocols can lead to significant variability in therapy outcomes, raising doubts about their overall efficacy (Vasser & Aru, 2020). To this day, one of the greatest challenges in psychology lies in the correct use of reproducibility and replicability (Legrenzi & Umiltà, 2023). Another challenge is motion sickness, a common problem many users experience during VR use. As reported by Wilson and Soranzo (2015), some researchers have noted physical and psychological side effects from VR exposure. These are collectively referred to as virtual reality-induced side effects (VRISE) (Sharples et al., 2008) and often focus on a general sense of malaise or motion sickness experienced by users immediately after a VR session (Murata, 2004). Initially, it was believed that the effect was caused by early VR technologies (low resolution and poor usability); a delay was often recorded between participant movements and display updates, resulting in a sense of disconnection between the user's perceptual and motor systems (Biocca, 1992). While technological advances have overcome this initial limitation, VRISE still remains and is an issue that needs further analysis (Keshavarz et al., 2013; Howarth & Hodder, 2008; Sugita et al., 2008). Specifically, motion sickness can manifest as dizziness, nausea, and disorientation, limiting the time individuals can spend in VR environments and negatively impacting the therapeutic experience. This issue is particularly relevant for patients who already suffer from balance disorders or other physical conditions that could be exacerbated by VR. It is also worth noting, as Riva (2022) recalls, that the first generation of VR devices (1990–2015) was characterized by low display resolution, limited field of view, and uncomfortable designs, which could consistently induce feelings of malaise.

Ecological Validity: this represents one of the most significant challenges of VR experiences. It is an assessment connected to the psychometric field but finds completeness only when applied practically. Ecological validity is closely tied to the individual's habitual behavioral repertoire and the environment in which they usually interact. Ecological validity, therefore, expresses the degree of "naturalness" of the test for the individual. Imagining ecological validity along a continuum, it is possible to identify at one end the maximum degree of ecology (a test extremely natural for the subject, requiring habitual behaviors in a daily life context) and at the other end the maximum degree of artificiality (a test entirely foreign to the subject's normal activities in a different environment from the usual one). It is important to understand that ecological validity must be evaluated on a case-by-case basis, considering the compromises given by the need to collect the required data without altering the individual's usual field of action. Indeed, it must be noted that the higher the ecological validity, the greater the possibility of generalizing the results obtained to the subject's daily life (Benatti & Zuin, 2017). Although VR simulations can create controlled and replicable environments, there is a risk that these experiences do not faithfully reflect real situations. This raises questions about the generalizability of results obtained in VR environments compared to the real world. Additionally, the concept of plasticity in the human mind should not be forgotten. This concept highlights how human behavior is highly susceptible to external influences, of which individuals are often unaware. This plasticity suggests that behavior is context-sensitive, and the mind is continuously shaped by various causal factors, including technology and other humans. These factors create unconscious influences on behavior, demonstrating that even seemingly stable character traits can change when the environmental context changes significantly. All these elements suggest that our environment, including technology, has an unconscious influence on our behavior (Madary & Metzinger, 2016). Furthermore, it should be noted that the ecological validity of behavioral studies in VR is not guaranteed even when immersive VR is used (Kulik, 2018).

Ethics: the ethical component concerning the use of VR in psychology raises particular difficulties in both experimental and clinical contexts. The immersive nature of VR can expose participants to excessive stress or traumatic experiences. It is essential to ensure that research protocols include safeguards to protect participants' well-being (Madary & Metzinger, 2016). Ethical considerations are fundamental in the use of VR in psychology. The protection of participants is the utmost priority, and rigorous measures must be adopted to ensure that VR experiences do not cause psychological harm. This includes the preventive assessment of risks, supervision during sessions, and the availability of psychological support post-experience. Another aspect related to the ethical component concerns informed consent. Participants must be fully aware of the potential risks and benefits of using VR. They must understand how their data will be used and have the option to withdraw from the experiment at any time without negative consequences. Another element that enriches and complicates the ethical approach is the aspect of justice and equity in access to VR technologies. It is important that these technologies are accessible not only to individuals and institutions with financial resources but also to disadvantaged groups who could benefit from VR research and clinical applications. Finally, as proposed by Sharma et al. (2014), a plausible way to make the virtual world ethically responsive is collective responsibility, which suggests that society has the power to influence but not control behavior in the virtual world.

Data Privacy: in connection with the ethical component, it is crucial to evaluate the ways in which individual data is managed. VR experiences can collect various types of data: sensitive data (personal information, etc.) related to the individual, emotional response patterns, movement patterns, biometric data (such as eye movements, heart rate, and physiological responses), and responses connected to participant interactions. These data can be used to create detailed user profiles, with significant privacy implications. Therefore, it is imperative to ensure data protection, block unauthorized access, and guarantee the use of this data solely for previously agreed-upon research or clinical/therapeutic purposes. Adopting robust data management practices, including encryption and strict data access policies, is essential to mitigate these risks (Madary & Metzinger, 2016). Specifically, various technological solutions can help protect the privacy of VR users. For instance, the use of end-to-end encryption can ensure that data collected during VR sessions is protected against malicious access. Encryption ensures that only authorized parties can decrypt and access sensitive data. Another solution is the implementation of differential privacy techniques. These techniques add noise to the collected data, making it more difficult for attackers to identify specific information about users. Differential privacy is particularly useful for protecting aggregated data used in research and user behavior analysis (Dwork & Roth, 2014). Lastly, user education and awareness play a critical role in privacy protection. Users need to be informed about the risks associated with the collection and use of their data and the measures they can take to effectively protect their privacy. Awareness campaigns and training programs can help users understand the importance of privacy and make informed decisions about VR use. Organizations should also promote a privacy culture among their employees and collaborators. This includes continuous training on data management practices, promoting stringent corporate privacy policies, and implementing regular and rigorous checks to verify compliance with regulations and ethical guidelines. The implications of privacy in VR applications are complex and require particular attention from all involved parties. Secure data collection and storage, anonymization, informed consent, regulatory compliance, and user education are all crucial elements to ensure that VR use is ethical and safe. As VR technology evolves, it will be essential to continue monitoring and addressing privacy challenges to protect users and maximize the therapeutic and research benefits of VR.

Technological Dependency Risk: the increasing use of virtual reality (VR) across various spheres of daily life, including clinical, therapeutic, educational, and entertainment contexts, highlights new aspects related to technological dependency that warrant attention. Prolonged use of VR raises concerns about addiction. VR addiction can stem from various factors, with the high level of immersion offered by VR being particularly significant. The highly realistic and engaging virtual environment can make it challenging to detach from the virtual experience. The allure of experiencing such an immersive situation can generate a gratifying experience, making it difficult for some individuals to leave the session, potentially leading to compulsive use (Barreda-Ángeles & Hartmann, 2022). Another factor to consider is the sense of escape that VR offers. Users can detach from reality and seek refuge in a "safe" virtual world. In this sense, users may use VR to escape real-life problems, finding comfort and gratification in virtual worlds where they have greater control and can experience successes and adventures that might not be accessible or possible in the real world. The results from Barreda-Ángeles and Hartmann (2022) indicate that VR applications do not present a higher risk of addiction compared to other traditional technologies. However, the embodiment sensations experienced during VR use are positively correlated with addiction. This might imply that future developments in VR technology could increase its addiction potential compared to other technologies. The potential development of traits associated with technological addiction can manifest in various ways. Common signs include excessive use of VR accompanied by difficulty limiting the time spent in VR and constant thoughts about the next opportunity to use it. Additionally, in aggravated situations, individuals may neglect daily responsibilities such as work, study, and interpersonal relationships in favor of time spent in VR (King et al., 2020). Another symptom related to addiction is tolerance, where there is systematic adaptation to stimuli and the related activation, necessitating the use of VR more frequently, for longer periods, and seeking increasingly intense experiences. Analyzing the issues expressed above makes it evident that technological addiction to VR can have physical and psychological consequences. Physically, prolonged use of VR can lead to issues such as eye strain, headaches, and musculoskeletal pain. Additionally, the lack of physical activity associated with prolonged VR use can contribute to long-term health problems like obesity and cardiovascular diseases (Stanney et al., 2020). Psychologically, VR addiction can worsen symptoms of pre-existing conditions like anxiety and depression. Individuals may also experience a reduced ability to cope with real-life problems, as they tend to retreat into the virtual world rather than face external challenges. Furthermore, VR addiction can incentivize isolation from social contexts, compromising interpersonal relationships and contributing to feelings of loneliness and alienation (Billieux et al., 2015). To effectively address technological addiction to VR, a multidisciplinary approach involving psychological, educational, and technological interventions is necessary. For instance, cognitive-behavioral therapy (CBT) can be effective in treating addiction by helping individuals identify and modify thoughts and behaviors contributing to excessive VR use (Young & De Abreu, 2015). Educational interventions are equally important. Educating users about the risks of technological addiction and promoting responsible VR use can raise awareness and encourage healthier behaviors. Useful activities can include awareness campaigns, workshops, and informative materials distributed in educational and community contexts (Young & De Abreu, 2015). Technological solutions can also play a significant role in mitigating addiction. For example, monitoring and control systems for usage time can help users maintain a healthy balance, limiting VR time and encouraging regular breaks. Additionally, VR designers can integrate elements that promote responsible use, such as time-spent notifications and suggestions for alternative activities (Bickmore & Picard, 2005).

Technoreligions: a final, but equally important, limitation lies in what are called technoreligions, or the absolute faith in new technologies as the solution to all human problems. The infinite possibilities offered by technology can make any problem seem easily solvable. This concept is well explored by Evgeny Morozov (2013), who critiques technological "solutionism," the idea that every problem can be solved with a technological solution. Blind faith in technology can lead to an overestimation of its capabilities, particularly regarding VR in psychology. The real risk is neglecting the limits and complexities of the human condition by relying solely on an absolute "belief" in new technologies. In the context of VR in psychology, this attitude can lead to an overestimation of the therapeutic capabilities of technology, overlooking its limitations and application issues. Blind faith in technology could also devalue human interactions and traditional therapeutic practices. Furthermore, the ease of access to VR devices (Vasser & Aru, 2020) combined with excessive trust (Morozov, 2013) in VR can create a dependence on technological solutions, neglecting the need to develop fundamental therapeutic and relational skills essential for the success of psychological treatment.

7.2   FUTURE CHALLENGES AND APPLICATIONS OF VR IN PSYCHOLOGY

Virtual reality is emerging as one of the most revolutionary and promising technologies of our time, with applications ranging from entertainment to clinical uses (medical and/or psychological), education, and professional training. It can therefore be connected to a myriad of application fields After consciously analyzing the limitations of VR technology, attention will now be focused on future perspectives and the challenges that virtual reality is set to face. Starting precisely from the aforementioned limitations, it becomes possible to generate a reflection on future applications. On one hand, the continuous evolution of technologies seems to overshadow the negative implications associated with the frantic pursuit of new technology. On the other hand, the developments in VR can encourage increasingly significant investments in the creation of more accessible and less expensive devices, consequently increasing the dissemination and adoption of these technologies on a large scale. Consider that eight years ago, in March 2016, the first generation of virtual reality headsets aimed at a broad consumer audience was released. The Oculus Rift, a head-mounted display (HMD) developed and produced by Oculus VR, a division of Facebook Inc., was sold at a starting price of 600 USD. It marked a new generation of VR devices, exponentially revolutionizing the use of VR. In a relatively short time, there has been a significant decrease in the cost of complete VR devices (input, output, and 3D graphical computation). We have moved from tens of thousands of dollars to a few hundred for the most affordable standalone VR systems (Riva, 2022).

7.2.1   Continuous Development and Pursuit of Immersiveness

Future applications aim to enhance user interface and immersive experiences, reducing negative side effects and thereby improving overall user experience, with evident implications for therapeutic efficacy. The goal is to make the experience increasingly detailed and personalized. In other words, VR technology seeks to predict the sensory consequences of users' actions, showing them the same expected outcomes their brains would anticipate in the real world (Riva, 2022). As highlighted by Riva et al. (2019), the VR system, much like the brain, maintains a model of the body and surrounding environment. This model is then used to anticipate and generate sensory inputs through VR hardware. To maximize the sense of presence, the VR model strives to faithfully mimic the brain model. Presence is therefore considered a crucial phenomenon for eliciting responses in VR similar to those in real life, yet it remains a concept difficult to objectively measure (Kober et al., 2012; Slater et al., 2009; Wilson & Soranzo, 2015; Vasser & Aru, 2020). Methods currently used to measure presence mainly consist of subjective self-assessment questionnaires, and many popular scales developed before the new generation of VR are unable to assess qualitative differences among various modern immersive techniques (Vasser & Aru, 2020).

The ability to create controlled and highly personalized research environments allows for the exploration of new research questions and testing hypotheses that would be challenging to investigate in the real world. For instance, VR can be used to study human behavior in high-stress situations, examine how people interact in complex social environments, or simulate scenarios that are difficult to replicate (Parsons, 2015).

Virtual reality can be described as a sophisticated system of imagination (Riva, 2022): an experiential form of fantasy that can be as powerful as reality itself in eliciting emotional responses. Studies such as those by North et al. (1997), Vincelli (1999), and Vincelli et al. (2001) demonstrate that VR can induce emotions and reactions with the same intensity and authenticity as real situations. VR thus deceives the predictive coding mechanisms that regulate bodily experience, generating the possibility of making people feel "real" in situations that are not (Riva, 2022).

As reported by Slater et al. (2020), through VR implementation, a person with light skin can temporarily have dark skin or vice versa (Maister et al., 2013, 2015; Peck et al., 2013; Banakou et al., 2016); an adult can become a child (Banakou et al., 2013; Tajadura-Jiménez et al., 2017); or it may be possible to experience a different height (taller or shorter) than in the real world (Yee & Bailenson, 2007; Freeman et al., 2014). Beyond changing or modifying certain bodily characteristics, VR allows for an infinite range of possible experiences from a first-person perspective. For example, one can be exposed to a virtual representation of a phobic agent, knowing it is not real but feeling as if it were. This has made VR increasingly utilized for therapeutic purposes, including pain management (Matamala-Gomez et al., 2019) and treatment of phobias and anxiety disorders (Freeman et al., 2017). The therapeutic potential in other areas has also been experimentally tested, such as for physical rehabilitation (Levin et al., 2015), rehabilitation of violent offenders (Seinfeld et al., 2018), and assessment of symptoms and neurocognitive deficits in individuals living with or at risk of psychosis (Rus-Calafell et al., 2018). VR enables the creation of new ways to structure, enhance, and/or replace bodily experience for clinical goals (Riva, 2008, 2016; Riva et al., 2016). Based on the examples provided, it is evident how VR applications can be advantageous in various experiential contexts, particularly offering advantages over in vivo exposure (Riva et al., 2015):

Cost: in vivo exposure is costly as it requires therapists to physically accompany patients to feared locations. While "therapist-free" exposure interventions are still uncommon, patients often hesitate to participate in such treatments.

Accessibility: feared situations are not always easily accessible, and imaginal exposure (i.e., exposure to imagined situations) is less effective in these cases.

Engagement: the immersion and interaction possibilities offered by VR enhance engagement compared to traditional interventions, thereby increasing subject participation.

Control: VR exposure allows almost complete control over what happens in the virtual world, including elements that can make situations more or less threatening (e.g., number or size of feared people, animals, or objects; height of spaces; presence of protective elements, etc.). Moreover, the therapist can monitor the situation, which elements the patient confronts, and what tends to disturb them or not. It's also possible to control the framing of the experience, as highlighted by Balzarotti and Ciceri (2014), where positively framed experiences tend to generate less fear than negatively framed ones.

Realism and Presence: unlike imaginal exposure, users undergoing VR feel present and judge their situation as real. This aspect is crucial for exposure therapy, as it aims to facilitate emotional processing of fear memories.

Beyond Reality: the use of virtual worlds allows the creation of situations or elements that are so "difficult or threatening" that one would not expect them to exist in the real world.

Personal Efficacy: VR is an important source of empowerment toward personal efficacy perception. It allows the construction of "virtual adventures" where individuals perceive themselves as competent and effective. The ultimate goal is to enable individuals to discover that obstacles and feared situations can be overcome through confrontation and continuous effort.

Safety: comparing in vivo and VR situations, the former can be highly aversive for patients and make them feel insecure, as there are no guarantees that something won't go wrong (e.g., elevator stops, technical issues on a plane, etc.). In contrast, VR's strength and advantage lie in its safety. In VR applications, patients can control the context and computer-generated environment to their liking, with no risk involved.

Privacy and Confidentiality: VR's ability to confront many fears in a therapist's controlled setting, without the need for in vivo exposure, offers significant advantages in terms of privacy and confidentiality.

7.2.2   *Innovations and Hardware Developments and Software Integration*

In addition to practical applications of VR, it's important to consider the intertwining of Hardware and Software, as it's from the combination of these two components that virtual experiences are realized. Virtual reality technology comprises a set of sophisticated technologies: a computer or mobile device with a capable graphics card for interactive 3D display, controllers, and a head-mounted display (HMD) incorporating one or more position trackers. These trackers detect the user's position and orientation, sending this information to the computer where images are updated and displayed in real time (Riva, 2022). Defining these elements allows for the identification of key technological components in a VR system (Parsons et al., 2017): input devices, output devices, and the simulated scenario (virtual environment).

Specifically, as reported by Riva (2022), there are various input devices used in a VR system. These include tracking devices like data gloves, head position sensors, integrated cameras, eye trackers, and many others. Simultaneously, there are also pointing devices such as six-degree-of-freedom mice, trackballs, joysticks, etc.

Output devices encompass all technologies providing computer-generated information continuously delivered to the user. While visual channels remain predominant in VR, advanced VR systems also offer auditory, olfactory, and haptic feedback.

The final component is the simulated scenario, the 3D virtual environment (VE) generated by the computer. VEs are designed for user exploration and interaction, allowing users to perform various actions/movements. Multi-user virtual environments (MUVEs) enable two or more users to share the same simulated scenario. MUVEs facilitate communication and interaction between users through avatars (customized graphical representations of individuals) controlled directly by users in real time. On the other hand, embodied virtual agents are graphical representations of individuals controlled by computer via artificial intelligence. Another important distinction to address is between immersive and non-immersive VR (Gaggioli et al., 2009).

VR technology is classified as immersive when it can sensorially separate the user from the physical world and replace sensory input with a computer-generated simulated scenario. Head-mounted displays (HMDs) are the most common immersive VR technology. They block any visual contact with the external world, and the display inside reproduces computer-generated images. Thanks to various sensors integrated into the HMD, the computer-generated image dynamically adjusts to different viewing positions (Riva, 2022).

A more advanced and expensive type of immersive VR is the Cave Automatic Virtual Environment (CAVE) (Cruz-Neira et al., 1992). A CAVE consists of a cubic space surrounding the user with images on walls (including floor and ceiling), displayed by a series of stereoscopic projectors. To see the 3D graphics projected in the CAVE, users wear 3D glasses. A motion capture system records the user's real-time position (free to move within the space) and adjusts the images to maintain the viewer's perspective continuously (Riva, 2022).

In contrast, non-immersive VR systems cannot completely occlude the user's field of view. They utilize standard high-resolution monitors (desktop or laptop screens) as output devices. Additionally, interactive capabilities are limited, for example, they do not allow movement tracking. Non-immersive virtual environments include 3D video games and desktop-based 3D modeling applications (Riva, 2022).

Hardware innovations focus on higher resolutions, wider fields of view, and faster refresh rates, significantly enhancing user experience quality. One of the areas where the most progress has been made is in VR hardware. Next-generation VR headsets offer higher resolutions, wider fields of view, and faster refresh rates, significantly enhancing user experience quality. Another significant innovation is represented by haptic feedback devices, providing realistic tactile sensations. Devices like haptic gloves and suits allow users to "feel" the virtual environment, enhancing interaction and immersion. Developments in this field have the potential to revolutionize VR applications in fields such as medicine or rehabilitation, where sensory feedback is crucial (Culbertson et al., 2018). Noteworthy are interaction techniques, a crucial area of innovation in VR. Gesture-based controls and eye-tracking are replacing traditional manual controllers, offering more natural and intuitive interaction. Eye-tracking, in particular, can enhance immersion and efficiency by allowing users to interact with the virtual environment simply by looking at objects (Koulieris et al., 2019).

Brain-computer interfaces (BCIs) represent another promising frontier. These technologies enable users to control the virtual environment with their thoughts, opening new possibilities for people with disabilities and improving VR accessibility (Lebedev & Nicolelis, 2006).

Simultaneously, software evolution dedicated to VR has seen significant improvements, with more powerful rendering engines and advanced algorithms enabling the creation of highly realistic virtual environments. As of 2024, Unity and Unreal Engine 5 remain among the most widely used game engines for VR content development, thanks to their ability to handle complex graphics and real-time interactions. Furthermore, within the software evolution process, the adoption of artificial intelligence and machine learning is observed, developed, and implemented through software. It is precisely through the support provided by AI integration on the software side that the modeling of VR scenarios, not only more immersive but also tailored to individual user characteristics, becomes possible (Vince, 2004; Riva, 2023).

7.2.3   Integration of Artificial Intelligence into VR

The rapid and significant evolution of artificial intelligence (AI) in recent years has led to its integration into virtual reality (VR). It is crucial to consider both the advantages and limitations of this new technology, as highlighted by Gigerenzer (2023), who emphasizes the importance of "staying intelligent in an intelligent system." This serves as a reminder to be aware of AI's potential while effectively managing and governing its applications, as any technology, no matter how "good," can become ineffective or even lead to negative effects if misused.

AI is revolutionizing the field of virtual reality (VR), making experiences more dynamic, personalized, and engaging (Bailenson, 2018; Riva, 2023). The integration of machine learning algorithms into VR allows for the adaptation of virtual environments to user actions and preferences in real time (Vince, 2004; Riva, 2023). This approach significantly enhances interactivity and immersion, creating unprecedented user experiences.

With AI, virtual environments can be dynamically adjusted based on user behavior. Machine learning algorithms analyze user actions and adjust various elements of the environment accordingly, such as game difficulty, object placement, and even virtual weather conditions. This adaptability creates a more personal and engaging experience where users feel integral to the virtual world. AI implemented in this manner enhances and makes the concept of "presence" even more effective (Sheridan, 1992; Biocca, 1999; Riva, 2022). Another significant contribution of AI to VR is the improvement of non-player character (NPC) behavior. Using advanced AI techniques, NPCs now have the ability to react more naturally and realistically to user actions, thereby enhancing interaction between users and NPCs. Instead of following predefined scripts, NPCs can adapt to new and unexpected situations, offering consistent and believable behavioral responses (Vallance, 2023). This not only improves immersion but also makes VR games or simulations more engaging and unpredictable. For example, Inworld AI is creating sophisticated NPCs capable of conducting real-time natural conversations with humans in VR and plans to interact with additional human modalities such as gestures, body language, and emotions (www.inworld.ai). AI is also transforming the quality of images in VR. Techniques such as predictive rendering and AI-based upscaling enable the generation of high-quality graphics without the need for extremely powerful hardware. Predictive rendering uses AI algorithms to predict and pre-render upcoming frames, reducing latency and improving visual fluidity. AI-based upscaling increases image resolution in real time, offering superior visual quality even on less powerful devices. In conclusion, the integration of AI into VR is advancing the field by making virtual experiences more adaptive, realistic, and immersive. However, careful consideration of ethical implications and effective management of AI technologies are essential to harnessing VR's full potential while mitigating risks and ensuring positive user experiences in virtual environments.

7.2.4   *Application of VR in Psychology: Research, Clinical Applications, and Education*

Within the field of psychology, the applications of virtual reality (VR) are increasingly being developed for research, clinical applications, and education. VR is finding new applications in various fields, ranging from pure research to clinical applications and educational contexts. In terms of research, as reported by Vasser and Aru (2020), the total number of articles published annually in peer-reviewed academic journals using the keyword "virtual reality" doubled from approximately 6000 to over 12,000 entries from 2016 to 2019, according to EBSCO Information Services. This surge in interest is understandable because VR holds enormous potential for both basic science and research (Pan & de Hamilton, 2018; Miller et al., 2019; Peeters, 2019) as well as therapeutic approaches (Rizzo & Koenig, 2017).

Vasser and Aru (2020) also note that despite the emergence of VR research in the 1990s, the recent spike in publications can be attributed to the release of various affordable and high-quality VR head-mounted displays (HMDs) (Pan & de Hamilton, 2018; Slater, 2018). Consumer-oriented VR systems such as HTC Vive and Oculus Rift have provided more researchers with the opportunity to conduct psychological experiments using immersive VR setups, integrating computers, HMDs, body tracking sensors, special interface devices, and 3D graphics (Rizzo & Koenig, 2017). In research, the aim is to technologically immerse users in virtual worlds, inducing a sense of presence, the illusion of being in the virtual world and behaving accordingly (Slater, 2018). Setting up a modern VR lab is relatively easy and cost-effective, even with minimal initial knowledge (Pan & de Hamilton, 2018; Kourtesis et al., 2019). However, it's important to note that not all immersive VR systems provide the same level of presence (Sinesio et al., 2019). For example, two studies both using 'immersive VR' could yield different results because the setups provided different levels of presence. Unfortunately, presence is difficult to quantify, which has significant implications for reliability and validity control (Vasser & Aru, 2020).

Regarding clinical applications of virtual reality (VR) in psychology, both the potential and current limitations identified in scientific trials are highlighted. VR has proven effective in various areas such as anxiety disorders, pain management, eating disorders, psychosis, addictions, autism, and sexual disorders. Despite significant progress, empirical results are often limited by methodological flaws in individual studies (Riva, 2022).

In clinical psychology, Norcross et al. (2013), during their study, interviewed 70 therapy experts regarding techniques they planned to implement and increase in usage over the next decade; virtual reality was ranked 4th out of 45 options, with other computer-supported methods occupying 4 of the top 5 positions (Rizzo & Bouchard, 2019). The trend is to increasingly integrate VR support into clinical approaches, supported by clinical outcomes. Recent meta-analyses (Riva et al., 2016, 2019) have assessed more than 53 systematic reviews and meta-analyses exploring the current use of VR in clinical psychology. The analyses showed VR's use in anxiety disorders, eating and weight disorders, and pain management, with long-term effects generalizing to the real world. VR also has potential applications in other areas such as psychosis and addictions (Riva, 2022).

A significant reflection has been developed by Rizzo and Bouchard (2019), who note how the recent proliferation of affordable HMDs and dedicated software for creating virtual environments may lead to the "misguided" belief that anyone with a good idea can create and market clinical tools. However, developing specific clinical tools without the guidance of scientific knowledge and clinical experience can lead to the production of ineffective or even dangerous applications for the subjects themselves (Rizzo et al., 2004).

Regarding educational impact, Foreman (2009) provides insightful reflections, indicating that it is not impossible to imagine that in 50 years, a significant portion of school and university material will be delivered through VR-integrated structures. Some undergraduate educational experiences might involve downloading virtual lectures, virtual teachers, and virtual seminar rooms (Foreman, 2009). In education, VR offers immersive learning experiences that can enhance understanding and retention of information. Students can explore reconstructed historical environments, conduct virtual experiments, and interact with educational materials in new and engaging ways (Merchant et al., 2014).

However, acquiring information in this manner does not represent the entire "learning experience," and many activities that constitute the "student experience," such as the social bond promoted by direct personal proximity, are probably irreplaceable (Foreman, 2009).

There are practical difficulties with virtual education that should not be underestimated, particularly because these apply to all subjects and not just psychology (Foreman, 2009). Firstly, to interact with materials (Merchant et al., 2014), there is a need to regularly update educational materials. This can be straightforward when only additional PowerPoint slides need to be inserted into a presentation but not when reprogramming a virtual lesson is required. Additionally, features such as emphasis, emotions, and spontaneous instructive digressions that characterize the lesson risk not being represented within a virtual presentation. In the context of learning, the progressive processing of knowledge through personal activity is a central element of current learning theories (Howe, 1998). It remains to be seen whether virtual environments can help provide such activity as part of distance learning programs. Unfortunately, technological developments in education have sometimes been considered mere gadgets and have not provided the originally anticipated benefits (Foreman, 2009). In conclusion, VR is proving to be a transformative technology in psychology, offering innovative solutions in research, clinical practice, and education. However, careful consideration of its limitations and ethical implications is essential to harnessing its full potential while ensuring positive outcomes for users in various contexts.

7.3   FINAL REFLECTIONS

Despite significant progress, virtual reality (VR) still faces several challenges before reaching its full potential. As observed, issues of accessibility and cost remain significant barriers, especially for users in developing countries. On the other hand, more affordable devices (Rizzo & Bouchard, 2019; Vasser & Aru, 2020; Riva, 2022) can facilitate the diffusion and utilization of VR technology. Smartphones and 360-degree videos will enable the development of a new generation of self-help applications in virtual reality, making mental health treatment more accessible to those who lack the time or economic resources to consult a clinician in person (Riva, 2022).

The increasing popularity of VR from scientific discoveries, social media, conferences, and innovative startups may give the impression that VR is something extremely new. However, it's important to remember that clinical VR applications were already proposed in the mid-1990s by many scholars (Lamson, Pugnetti, Rothbaum, Riva, Rizzo, Weiss, and Wiederhold, cited in alphabetical order), among others. Moreover, various scientific journals, conferences, and manuals dedicated to the subject continue to report scientific results for decades (Rizzo & Bouchard, 2019). The tradition of research, VR application, and its growing dissemination highlight the importance of developing increasingly specific and detailed ethical approaches and privacy management. Ethical issues, including privacy management and potential technological dependence, require attention. There is a need for ethical awareness in VR use and accurate training regarding data management and privacy. The collection of personal data through VR devices raises concerns about privacy protection and the ethical use of such information (Madary & Metzinger, 2016; Stanney et al., 2020). Building on the work of Madary and Metzinger (2016), effective guidelines can be outlined: (1) No experiment should be conducted using virtual reality if serious or lasting harm to a subject is foreseeable; (2) Informed consent is recommended for VR experiments and should include an explicit statement that immersive VR may have lasting behavioral influences on subjects and that some of these risks may currently be unknown; (3) Researchers dealing with VR with the aim of developing new clinical applications should proceed slowly and cautiously, in close collaboration with physicians, to make informed judgments about the suitability of particular patients for new trials; (4) In general, scientists and the media must be clear and honest with the public about scientific progress, especially in the area of VR use for medical treatment; (5) Torture in a virtual environment is still torture. The fact that suffering occurs while immersed in a virtual environment does not diminish the suffering itself; (6) Urgent call for the scientific community to take steps to prevent the abuse of informed consent with this technology, especially in the interest of preserving public trust; (7) Scientists must understand that following an ethical code is not the same as being ethical. A specific ethical code for a domain, however consistent, developed, and detailed it may be in future versions, can never function as a substitute for ethical reasoning itself. Therefore, it is important to discuss the implications of these technologies for brain organization and reality perception, highlighting the need to balance technological advancements with deep ethical considerations (De Pisapia, 2024).

Additionally, dependence on VR can have negative consequences on users' mental and physical health, necessitating the promotion of responsible usage practices (Billieux et al., 2015). The ecological validity of VR experiences is another concern, as virtual environments may not always faithfully replicate real-world conditions. Furthermore, comparing experimental dynamics proposed with different VR devices and settings is not always possible or equally comparable (Slater et al., 2009; Vasser & Aru, 2020).

As reported by Riva (2022), VR can also be considered an embodied technology, derived from its ability to deceive the brain mechanisms that regulate the body experience. This capacity allows the creation of new ways, hitherto only partially explored, to structure, enhance, and/or replace body experience for clinical goals. These new modes are incorporated to assess brain function, allowing the evaluation of processes behind behaviors observed in the real world. These features make VR an extremely effective clinical tool. Drawing on the results of two recent meta-reviews (Riva et al., 2016, 2019), existing research supports the clinical use of VR in the assessment and treatment of anxiety disorders, pain management, and eating and weight disorders, with long-term effects that can generalize to the real world. Moreover, recent studies have also provided preliminary support for the use of VR in the assessment and treatment of psychosis, addictions, and autism (Riva, 2022).

Moving toward final observations and attempting to unify all reflections made within this chapter, it is possible to state that innovations in virtual reality are transforming how we interact with technology, offering new opportunities and addressing significant challenges. From advancements in hardware and software to the integration of artificial intelligence and new applications, VR is continuously evolving and expanding its boundaries. These developments not only improve user experience but also enhance the simulation capability of complex scenarios for education, professional training, and therapies, both clinical/psychological and medical. VR is revolutionizing sectors such as education, enabling students to use immersive learning environments, and healthcare, offering new treatment modalities for psychological disorders (stress, anxiety, depression, and many more).

With a mindful and responsible approach, these innovations have the potential to bring significant benefits to society, improving quality of life and opening new frontiers in research and professional practice. However, it is essential to consider the ethical and social implications of widespread VR use. Protecting privacy, preventing abuses, and managing psychological effects must be top priorities to ensure that the technology, applied across various fields, is used safely and advantageously for all. Furthermore, the accessibility of VR technologies must be promoted to avoid a digital divide that could exclude less affluent segments of the population. The attempt to democratize access to VR technology must be an actively pursued goal to provide support to all segments of the population. Ultimately, the future of virtual reality depends on the ability of the human population to balance innovation and responsibility. With collaboration among developers, researchers, psychologists, physicians, and legislators, it is possible to create a VR ecosystem that not only controls and supervises the limits of technology but does so ethically and inclusively, ensuring that benefits are equitably distributed and risks are minimized.

References

Bailenson, J. (2018). *Experience on demand: What virtual reality is, how it works, and what it can do.* W. W. Norton & Company.

Balzarotti, S., & Ciceri, M. R. (2014). News reports of catastrophes and viewers' fear: Threat appraisal of positively versus negatively framed events. *Media Psychology,* 17(4), 357–377. https://doi.org/10.1080/15213269.2013.826588

Banakou, D., Groten, R., & Slater, M. (2013). Illusory ownership of a virtual child body causes overestimation of object sizes and implicit attitude changes. *Proceedings of the National Academy of Sciences of the United States of America,* 110(31), 12846–12851. https://doi.org/10.1073/pnas.1306779110

Banakou, D., Hanumanthu, P. D., & Slater, M. (2016). Virtual embodiment of white people in a black virtual body leads to a sustained reduction in their implicit racial bias. *Frontiers in Human Neuroscience, 10.* https://doi.org/10.3389/fnhum.2016.00601

Barreda-Ángeles, M., & Hartmann, T. (2022). Hooked on the metaverse? Exploring the prevalence of addiction to virtual reality applications. *Frontiers in Virtual Reality, 3,* 1031697. https://doi.org/10.3389/frvir.2022.1031697

Benatti, F., & Zuin, M. (2017). Elaborazione e requisiti delle prove psicodiagnostiche, Test 1. libreriauniversitaria.it edizioni.

Bickmore, T. W., & Picard, R. W. (2005). Establishing and maintaining long-term human-computer relationships. *ACM Transactions on Computer-Human Interaction,* 12(2), 293–327. https://doi.org/10.1145/1067860.1067867

Billieux, J., Thorens, G., Khazaal, Y., Zullino, D., Achab, S., & Van der Linden, M. (2015). Problematic involvement in online games: A cluster analytic approach. *Computers in Human Behavior, 43,* 242–250. https://doi.org/10.1016/j.chb.2014.10.055

Biocca, F. (1992). Will simulation sickness slow down the diffusion of virtual environment technology? *Presence: Teleoperators & Virtual Environments, 1,* 334–343. https://doi.org/10.1162/pres.1992.1.3.334

Biocca, F. (1999). Chapter 6 the Cyborg's dilemma. Progressive embodiment in virtual environments. In *Human factors in information technology* (Vol. 13 C, pp. 113–144). Elsevier. https://doi.org/10.1016/S0923-8433(99)80011-2

Cruz-Neira, C., Sandin, D. J., DeFanti, T. A., Kenyon, R. V., & Hart, J. C. (1992). The CAVE: audio visual experience automatic virtual environment. *Communications of the ACM,* 35(6), 64–72. https://doi.org/10.1145/129888.129892

Culbertson, H., Schorr, S. B., & Okamura, A. M. (2018). Haptics: The present and future of artificial touch sensation. *Annual Review of Control, Robotics, and Autonomous Systems,* 1(1), 385–409. https://doi.org/10.1146/annurev-control-060117-105043

De Pisapia, N. (2024). Understanding consciousness in the age of AI and XR: Altered states, emerging realities, and the digital self. In F. Santoianni, G. Giannini, & A. Ciasullo (Eds.), *Mind, body, and digital brains. Integrated science* (Vol. 20). Springer. https://doi.org/10.1007/978-3-031-58363-6_6

Dwork, C., & Roth, A. (2014). The algorithmic foundations of differential privacy. *Foundations and Trends® in Theoretical Computer Science, 9*(3–4), 211–407. https://doi.org/10.1561/0400000042

Foreman, N. (2009). Virtual reality in psychology. *Themes in Science and Technology Education, 2*(1), 225–252.

Freeman, D., Evans, N., Lister, R., Antley, A., Dunn, G., & Slater, M. (2014). Height, social comparison, and paranoia: An immersive virtual reality experimental study. *Psychiatry Research, 218*(3), 348–352. https://doi.org/10.1016/j.psychres.2013.12.014

Freeman, D., Reeve, S., Robinson, A., Ehlers, A., Clark, D., Spanlang, B., & Slater, M. (2017). Virtual reality in the assessment, understanding, and treatment of mental health disorders. *Psychological Medicine, 47*(14), 2393–2400. https://doi.org/10.1017/S003329171700040X

Gaggioli, A., Keshner, E. A., Weiss, P. L., & Riva, G. (Eds.). (2009). *Advanced technologies in rehabilitation - empowering cognitive, physical, social and communicative skills through virtual reality, robots, wearable systems and brain-computer interfaces.* IOS Press.

Gigerenzer, G. (2023). *Perché l'intelligenza umana batte ancora gli algoritmi.* Raffaello Cortina Editore.

Howarth, P. A., & Hodder, S. G. (2008). Characteristics of habituation to motion in a virtual environment. *Displays, 29*(2), 117–123. https://doi.org/10.1016/j.displa.2007.09.009

Howe, C. (1998). Psychology teaching in the 21st century. *The Psychologist, 11*(8), 371–390.

Inworld. (2024). *Inworld introduction.* Retrieved June 30, 2024, from https://docs.inworld.ai/docs/intro

Keshavarz, B., Hettinger, L. J., Vena, D., & Campos, J. L. (2013). Combined effects of auditory and visual cues on the perception of vection. *Experimental Brain Research, 232*(3), 827–836. https://doi.org/10.1007/s00221-013-3793-9

King, D. L., Delfabbro, P. H., Billieux, J., & Potenza, M. N. (2020). Problematic online gaming and the COVID-19 pandemic. *Journal of Behavioral Addictions, 9*(2), 184–186. https://doi.org/10.1556/2006.2020.00016

Kober, S. E., Kurzmann, J., & Neuper, C. (2012). Cortical correlate of spatial presence in 2D and 3D interactive virtual reality: An EEG study. *International Journal of Psychophysiology, 83*(3), 365–374. https://doi.org/10.1016/j.ijpsycho.2011.12.003

Koulieris, G. A., Akşit, K., Stengel, M., Mantiuk, R. K., Mania, K., & Richardt, C. (2019). Near-eye display and tracking technologies for virtual and augmented reality. *Computer Graphics Forum, 38*(2), 493–519. https://doi.org/10.1111/cgf.13654

Kourtesis, P., Collina, S., Doumas, L. A. A., & MacPherson, S. E. (2019). Technological competence is a pre-condition for effective implementation of virtual reality head mounted displays in human neuroscience: A technological review and meta-analysis. *Frontiers in Human Neuroscience, 13.* https://doi.org/10.3389/fnhum.2019.00342

Kulik, A. (2018). Virtually the ultimate research lab. *British Journal of Psychology, 109*(3), 434–436. https://doi.org/10.1111/bjop.12307

Lebedev, M. A., & Nicolelis, M. A. (2006). Brain-machine interfaces: Past, present and future. *Trends in Neurosciences, 29*(9), 536–546. https://doi.org/10.1016/j.tins.2006.07.004

Legrenzi, P., & Umiltà, C. (2023). *Il sapere come mestiere.* Il Mulino.

Levin, M. F., Weiss, P. L., & Keshner, E. A. (2015). Emergence of virtual reality as a tool for upper limb rehabilitation: Incorporation of motor control and motor learning principles. *Physical Therapy, 95*(3), 415–425. https://doi.org/10.2522/ptj.20130579

Madary, M., & Metzinger, T. K. (2016). Recommendations for good scientific practice and the consumers of VR-technology. *Frontiers in Robotics and AI, 3.* https://doi.org/10.3389/frobt.2016.00003

Maister, L., Sebanz, N., Knoblich, G., & Tsakiris, M. (2013). Experiencing ownership over a dark-skinned body reduces implicit racial bias. *Cognition, 128*(2), 170–178. https://doi.org/10.1016/j.cognition.2013.04.002

Maister, L., Slater, M., Sanchez-Vives, M. V., & Tsakiris, M. (2015). Changing bodies changes minds: Owning another body affects social cognition. *Trends in Cognitive Sciences, 19*(1), 6–12. https://doi.org/10.1016/j.tics.2014.11.001

Matamala-Gomez, M., Donegan, T., Bottiroli, S., Sandrini, G., Sanchez-Vives, M. V., & Tassorelli, C. (2019). Immersive virtual reality and virtual embodiment for pain relief. *Frontiers in Human Neuroscience, 13.* https://doi.org/10.3389/fnhum.2019.00279

Merchant, Z., Goetz, E. T., Cifuentes, L., Keeney-Kennicutt, W., & Davis, T. J. (2014). Effectiveness of virtual reality-based instruction on students' learning outcomes in K-12 and higher education: A meta-analysis. *Computers & Education, 70,* 29–40. https://doi.org/10.1016/j.compedu.2013.07.033

Miller, L. C., Shaikh, S. J., Jeong, D. C., Wang, L., Gillig, T. K., Godoy, C. G., Appleby, P. R., Corsbie-Massay, C. L., Marsella, S., Christensen, J. L., & Read, S. J. (2019). Causal inference in generalizable environments: Systematic representative design. *Psychological Inquiry, 30*(4), 173–202. https://doi.org/10.1080/1047840x.2019.1693866

Morozov, E. (2013). *To save everything, click here: The folly of technological solutionism.* Public Affairs.

Murata, A. (2004). Effects of duration of immersion in a virtual reality environment on postural stability. *International Journal of Human-Computer Interaction, 17*(4), 463–477. https://doi.org/10.1207/s15327590ijhc1704_2

Norcross, J. C., Pfund, R. A., & Prochaska, J. O. (2013). Psychotherapy in 2022: A Delphi poll on its future. *Professional Psychology: Research and Practice, 44*(5), 363–370. https://doi.org/10.1037/a0034633

North, M. M., North, S. M., & Coble, J. R. (1997). Virtual reality therapy: An effective treatment for psychological disorders. *Studies in Health Technology and Informatics, 44,* 59–70.

Pan, X., & de Hamilton, A. F. C. (2018). Why and how to use virtual reality to study human social interaction: The challenges of exploring a new research landscape. *British Journal of Psychology, 109*(3), 395–417. https://doi.org/10.1111/bjop.12290

Parsons, T. D. (2015). Virtual reality for enhanced ecological validity and experimental control in the clinical, affective and social neurosciences. *Frontiers in Human Neuroscience.* https://doi.org/10.3389/fnhum.2015.00660

Parsons, T. D., Riva, G., Parsons, S., Mantovani, F., Newbutt, N., Lin, L., Venturini, E., & Hall, T. (2017). Virtual reality in pediatric psychology. *Pediatrics, 140*(Suppl 2), S86–S91. https://doi.org/10.1542/peds.2016-1758I

Peck, T. C., Seinfeld, S., Aglioti, S. M., & Slater, M. (2013). Putting yourself in the skin of a black avatar reduces implicit racial bias. *Consciousness and Cognition, 22*(3), 779–787. https://doi.org/10.1016/j.concog.2013.04.016

Peeters, D. (2019). Virtual reality: A game-changing method for the language sciences. *Psychonomic Bulletin & Review.* https://doi.org/10.3758/s13423-019-01571-3

Riva, G. (2008). From virtual to real body: Virtual reality as embodied technology. *Journal of Cybertherapy and Rehabilitation, 1*(1), 7–22.

Riva, G. (2016). Embodied medicine: What human-computer confluence can offer to health care. In A. Gaggioli, A. Ferscha, G. Riva, S. Dunne, & I. Viaud-Delmon (Eds.), *Human computer confluence: Transforming human experience through symbiotic technologies.* De Gruyter Open.

Riva, G. (2022). Virtual reality in clinical psychology. *Comprehensive Clinical Psychology,* 91–105. https://doi.org/10.1016/B978-0-12-818697-8.00006-6

Riva, G. (2023). *Virtual reality.* Springer.

Riva, G., Baños, R. M., Botella, C., Mantovani, F., & Gaggioli, A. (2016). Transforming experience: The potential of augmented reality and virtual reality for enhancing personal and clinical change. *Frontiers in Psychiatry, 7,* 164. https://doi.org/10.3389/fpsyt.2016.00164

Riva, G., Wiederhold, B. K., & Mantovani, F. (2019). Neuroscience of virtual reality: From virtual exposure to embodied medicine. *Cyberpsychology, Behavior and Social Networking, 22*(1), 82–96. https://doi.org/10.1089/cyber.2017.29099.gri

Riva, G., et al. (2015). Presence-inducing Media for Mental Health Applications. In M. Lombard, F. Biocca, J. Freeman, W. IJsselsteijn, & R. Schaevitz (Eds.), *Immersed in media.* Springer. https://doi.org/10.1007/978-3-319-10190-3_12

Rizzo, A. "Skip", & Bouchard, S (Ed.). (2019). Virtual reality for psychological and neurocognitive interventions. *Virtual reality technologies for health and clinical applications*. https://doi.org/10.1007/978-1-4939-9482-3

Rizzo, A. A., & Koenig, S. T. (2017). Is clinical virtual reality ready for primetime? *Neuropsychology, 31*(8), 877–899. https://doi.org/10.1037/neu0000405

Rizzo, A. A., Schultheis, M., Kerns, K. A., & Mateer, C. (2004). Analysis of assets for virtual reality applications in neuropsychology. *Neuropsychological Rehabilitation, 14*(1–2), 207–239. https://doi.org/10.1080/09602010343000183

Rus-Calafell, M., Garety, P., Sason, E., Craig, T. J. K., & Valmaggia, L. R. (2018). Virtual reality in the assessment and treatment of psychosis: A systematic review of its utility, acceptability and effectiveness. *Psychological Medicine, 48*(3), 362–391. https://doi.org/10.1017/S0033291717001945

Seinfeld, S., Arroyo-Palacios, J., Iruretagoyena, G., Hortensius, R., Zapata, L. E., Borland, D., de Gelder, B., Slater, M., & Sanchez-Vives, M. V. (2018). Offenders become the victim in virtual reality: Impact of changing perspective in domestic violence. *Scientific Reports, 8*(1), 2692. https://doi.org/10.1038/s41598-018-19987-7

Sharma, S., Jerripothula, S., Mackey, S., & Soumare, O. (2014). Immersive virtual reality environment of a subway evacuation on a cloud for disaster preparedness and response training. *2014 IEEE symposium on computational intelligence for human-like intelligence (CIHLI)*. https://doi.org/10.1109/cihli.2014.7013380

Sharples, S., Cobb, S., Moody, A., & Wilson, J. (2008). Virtual reality induced symptoms and effects (VRISE): Comparison of head mounted display (HMD), desktop and projection display systems. *Displays, 29*(2), 58–69. https://doi.org/10.1016/j.displa.2007.09.005

Sheridan, T. B. (1992). Musings on telepresence and virtual presence. *Presence: Teleoperators and Virtual Environments, 1*(1), 120–126. https://doi.org/10.1162/pres.1992.1.1.120

Sinesio, F., Moneta, E., Porcherot, C., Abbà, S., Dreyfuss, L., Guillamet, K., Bruyninckx, S., Laporte, C., Henneberg, S., & McEwan, J. A. (2019). Do immersive techniques help to capture consumer reality? *Food Quality and Preference*. https://doi.org/10.1016/j.foodqual.2019.05.004

Slater, M. (2018). Immersion and the illusion of presence in virtual reality. *British Journal of Psychology, 109*(3), 431–433. https://doi.org/10.1111/bjop.12305

Slater, M., Gonzalez-Liencres, C., Haggard, P., et al. (2020). The ethics of realism in virtual and augmented reality. *Frontiers in Virtual Reality, 1*. https://doi.org/10.3389/frvir.2020.00001

Slater, M., Lotto, B., Arnold, M. M., & Sanchez-Vives, M. V. (2009). How we experience immersive virtual environments: The concept of presence and its measurement. *Anuario de Psicología, 40*(2), 193–210.

Stanney, K., Lawson, B. D., Rockers, B., et al. (2020). Identifying causes of and solutions for cybersickness in immersive technology: Reformulation of a research and development agenda. *International Journal of Human–Computer Interaction*, *36*(19), 1783–1803. https://doi.org/10.1080/1044731 8.2020.1828535

Sugita, N., Yoshizawa, M., Tanaka, A., Abe, K., Chiba, S., Yambe, T., & Nitta, S. (2008). Quantitative evaluation of effects of visually-induced motion sickness based on causal coherence functions between blood pressure and heart rate. *Displays*, *29*(2), 167–175. https://doi.org/10.1016/j.displa. 2007.09.017

Tajadura-Jiménez, A., Banakou, D., Bianchi-Berthouze, N., & Slater, M. (2017). Embodiment in a child-like talking virtual body influences object size perception, self-identification, and subsequent real speaking. *Scientific Reports*, *7*(1), 9637. https://doi.org/10.1038/s41598-017-09497-3

Vallance, M. (2023). Independently supporting learners in VR with an AI-enabled non-player character (NPC). *Immersive Learning Research - Practitioner*, *1*(1), 69–73.

Vasser, M., & Aru, J. (2020). Guidelines for immersive virtual reality in psychological research. *Current Opinion in Psychology*, *36*, 71–76. https://doi. org/10.1016/j.copsyc.2020.04.010

Vince, J. (2004). *Introduction to virtual reality*. Springer.

Vincelli, F. (1999). From imagination to virtual reality: The future of clinical psychology. *Cyberpsychology & Behavior*, *2*(3), 241–248. https://doi. org/10.1089/109493199316366

Vincelli, F., Molinari, E., & Riva, G. (2001). Virtual reality as clinical tool: Immersion and three-dimensionality in the relationship between patient and therapist. *Studies in Health Technology and Informatics*, *81*, 551–553.

Wilson, C. J., & Soranzo, A. (2015). The use of virtual reality in psychology: A case study in visual perception. *Computational and Mathematical Methods in Medicine*. https://doi.org/10.1155/2015/151702

Yee, N., & Bailenson, J. (2007). The proteus effect: The effect of transformed self-representation on behavior. *Human Communication Research*, *33*(3), 271–290. https://doi.org/10.1111/j.1468-2958.2007.00299.x

Young, K. S., & De Abreu, C. N. (2015). *Internet addiction: A handbook and guide to evaluation and treatment*. Wiley.

CHAPTER 8

Conclusions

Abstract Virtual Reality (VR) has significantly transformed psychological research, treatment, and education. This paper reviews major discoveries and innovations facilitated by VR, highlighting its efficacy in treating anxiety, PTSD, and phobias through controlled, realistic simulations. VR's ability to create immersive environments allows for gradual desensitization in a safe context, offering new therapeutic possibilities. Notably, VR-based exposure therapy has demonstrated effectiveness in reducing symptoms of acrophobia and PTSD in war veterans. Beyond clinical applications, VR enhances educational outcomes by providing multisensory, interactive learning environments that improve student engagement and information retention. VR also plays a pivotal role in simulating complex clinical scenarios for training future mental health professionals. Furthermore, VR's integration with mindfulness techniques has shown promising results in managing grief and emotional loss, promoting psychological well-being through immersive, calming environments. Technological advancements, including AI and haptic feedback, have made VR more accessible and realistic, enhancing its therapeutic potential. Despite challenges related to accessibility, standardization, and ethical considerations, VR offers a powerful tool for advancing psychological practice and research. Future exploration of VR's integration with other therapeutic modalities and continuous technological evolution hold immense potential for improving mental health treatments and understanding cognitive and emotional processes.

Keywords Virtual reality • Mindfulness • Cyberpsychology • Clinical applications • VR environments

8.1   Summary of Major Discoveries and Innovations Through Virtual Reality (VR)

Virtual Reality (VR) has opened new frontiers in psychology, expanding treatment and research capabilities in previously unimaginable ways. Innovations brought by VR in the psychological field include the effective treatment of disorders such as anxiety, PTSD, and phobias through controlled virtual environments that replicate realistic scenarios without associated risks. These environments allow experiencing and managing stressful situations in a safe context, thus facilitating gradual and controlled desensitization.

The effectiveness of VR in treating anxiety disorders and phobias is well documented. A notable example is the use of VR to treat acrophobia, where patients are gradually exposed to high-altitude situations in a controlled environment, improving their ability to manage anxiety in real-life situations (Emmelkamp et al., 2002). Similarly, VR has proven effective in treating PTSD in war veterans, with studies showing significant symptom reduction through simulated combat scenarios (Rothbaum et al., 1995).

Beyond clinical application, VR has shown great potential in education. It has improved learning outcomes by creating multisensory educational environments that stimulate student interaction and immersion. This experiential learning contributes to greater engagement and better information retention, showing a significant impact on training future professionals in psychology and mental health. Simulating complex clinical situations, such as diagnostic interviews or critical interventions, allows students to practice in a safe environment and receive immediate feedback, enhancing their practical skills (Garcia-Ruiz et al., 2008).

VR has revolutionized the field of psychology and personal well-being with significant discoveries and innovations. The use of VR to create immersive experiences that help reduce stress and mental fatigue has transformed the approach to psychological recovery. Recent studies, such as those conducted by Moyle et al. (2018) and Argüero-Fonseca et al. (2023), have demonstrated how VR can induce deep relaxation states and improve overall psychological well-being. This tool has been particularly effective in simulating natural environments that facilitate cognitive recovery, for example, improving the quality of life for patients with dementia, thus extending therapeutic applications beyond entertainment.

Simulating natural environments, such as forests or beaches, has been shown to have a significant calming effect on users, contributing to cortisol level reduction and mood improvement (Anderson et al., 2017). These virtual environments can be used not only to reduce stress but also to facilitate cognitive recovery in patients with neurological deficits, offering a rich and engaging sensory stimulus.

The adoption of VR has also enhanced the ecological validity of cognitive and emotional studies, highlighting how VR can evoke authentic emotional responses (Marchioro & Benatti, 2022; Marchioro et al., 2023). The ability to create realistic and interactive scenarios allows for more accurate study of cognitive and behavioral reactions compared to traditional laboratory methods. For instance, VR has been used to study stress responses in simulated work environments, enabling researchers to observe behaviors that would be difficult to replicate in a controlled context (Chen et al., 2024).

Furthermore, the ability to create dynamic and personalized learning environments has revolutionized the educational approach, significantly improving motivation among children and providing immediate feedback (Parsons & Cobb, 2011; Freina & Ott, 2015). The gamification of virtual educational environments has been shown to increase engagement and information retention, making learning more interactive and enjoyable. This approach has shown positive results not only in traditional education but also in training mental health professionals, where practicing realistic clinical scenarios is crucial.

Another innovation concerns the integration of VR with mindfulness, especially in the treatment of grief and emotional loss. VR allows the simulation of serene and safe environments that promote emotional processing and healing, as demonstrated by therapeutic experiences in virtual environments like forests or beaches (Riva et al., 2019; Botella et al., 2017). Personalized interventions in virtual environments have shown greater therapeutic effectiveness compared to traditional methods due to the ability to tailor scenarios to the specific needs of each individual (Maples-Keller et al., 2017). The combination of VR and mindfulness has shown promising results in reducing symptoms of anxiety and depression, improving overall patient well-being (Bailenson, 2018).

From a technological standpoint, advancements in hardware and software have made VR more accessible and of higher quality. Integration with artificial intelligence and machine learning algorithms has allowed for the creation of more dynamic and personalized virtual environments (Riva, 2022). Advanced sensory experiences through haptic feedback and predictive rendering techniques have further improved interaction and immersion, making virtual experiences more realistic and engaging (Culbertson et al., 2018). The use of haptic feedback, which provides tactile sensations through gloves or other devices, has increased the sense of presence and interactivity of virtual environments, making the experience more immersive and realistic.

The innovations introduced by virtual reality in the field of psychology have thus transformed both clinical practice and research. The ability to create controlled and realistic environments offers new opportunities for treating and studying psychological disorders, improving the effectiveness of therapies and the quality of data collected. With continuous technological advancement, VR is set to become an increasingly important component of psychological practice, opening new frontiers in the treatment and understanding of cognitive and emotional processes.

8.2   THE GROWING ROLE OF VR IN PSYCHOLOGY

The use of VR in psychology is growing exponentially, transforming not only the methods of treating mental disorders but also research methods and educational strategies in the field. The ability to simulate complex and controllable environments offers researchers the opportunity to study cognitive and emotional dynamics in scenarios that would otherwise be difficult or impossible to replicate in reality. This paves the way for a deeper understanding of psychological disorders and human behaviors.

The use of VR in psychological research has overcome many of the limitations of traditional methods. For example, VR allows for the creation of controlled and reproducible stress situations, which are fundamental for studying behavioral and physiological responses without exposing participants to real dangers. This rigorous control of experimental variables significantly improves the internal validity of studies (Slater, 2009).

In the clinical field, VR allows for innovation in intervention techniques by offering exposure therapies in environments that reduce patient anxiety, facilitating faster and more sustainable recovery. For example, VR-based exposure therapy for treating specific phobias has proven effective in significantly reducing symptoms compared to traditional techniques

(Botella et al., 2015). Moreover, the use of VR can help overcome some of the logistical and practical barriers of traditional therapies, such as the difficulty of accessing specialist treatments for those living in remote areas. This is particularly important to ensure equitable access to psychological care.

In the field of psychology, VR has acquired an increasingly central role as a complementary therapeutic tool. The ability to simulate environments that would otherwise be inaccessible makes VR a unique tool for treating various psychological disorders. For example, it has been used to help individuals with dementia through immersion in calming virtual environments that stimulate memory and cognitive engagement. In these cases, the creation of familiar environments or places significant to the patient has been shown to improve well-being and reduce some typical symptoms of dementia (Moreno et al., 2019).

Additionally, VR has shown remarkable results in reducing demotivation in children through exposure to playful scenarios that stimulate their interaction and participation. These advancements indicate how VR is expanding the field of therapeutic possibilities by offering non-invasive methods that complement traditional therapies and improve patients' quality of life. Integrating elements of play and challenge within therapeutic sessions helps maintain a high motivation and engagement in children, making therapy more effective (Lau et al., 2017).

The ability of VR to reproduce potentially dangerous or inaccessible contexts in complete safety has facilitated the treatment of conditions such as phobias and anxiety disorders (Parsons, 2015). For example, simulating flying in an airplane for those afraid of flying, or public speaking for those with social phobia, allows for gradual exposure that helps reduce anxiety and improve stress management skills. Furthermore, VR offers an unprecedented laboratory for observing how the human brain interacts with simulated realities, providing a deeper understanding of cognitive and emotional processes. Neuroimaging studies during VR use have shown how certain areas of the brain activate in response to virtual stimuli, providing new insights into the neural mechanisms underlying emotions and behaviors (Garrett et al., 2017).

The therapeutic applications of VR, such as in treating PTSD and anxiety disorders, are revolutionizing therapeutic practices by offering new possibilities for more effective treatments (Turner & Casey, 2014). The ability of VR to simulate complex experiences and elicit authentic emotional responses allows researchers to explore cognitive and emotional dynamics in ways previously inaccessible with traditional methods. This technology has shown promising results in treating PTSD, with significant reductions in traumatic memories and anxiety symptoms (Rothbaum et al., 1995).

In the psychological field, VR is emerging as an effective tool for engaging children in safe and stimulating therapeutic environments. For example, VR can simulate social scenarios for children with autism spectrum disorders, providing a controlled environment to practice social interactions without the stress of real-life consequences (Maskey et al., 2014). The interactive and customizable nature of VR allows for the creation of varied and engaging therapeutic exercises, maintaining high motivation and commitment of children in therapeutic programs. This approach is particularly useful for improving social skills and reducing problematic behaviors in children with autism (Parsons et al., 2017).

VR has also demonstrated the ability to overcome some of the limitations of traditional methods in grief therapy, such as reluctance to verbalize emotions and variability in therapeutic outcomes (Neimeyer, 2001; Jordan & Neimeyer, 2003). The ability to create immersive environments facilitates a deeper emotional connection, making interventions more effective. The possibility of reliving significant moments or interacting with virtual representations of deceased family members has been shown to facilitate the grieving process and improve emotional well-being (Riva et al., 2016).

Mindfulness programs through VR have demonstrated a significant reduction in symptoms of depression, anxiety, and perceived stress (Bailenson, 2018). These results suggest that VR can be effectively used to integrate mindfulness practices in grief therapy, offering more robust and immediate emotional support. The ability to immerse oneself in calming and relaxing virtual environments facilitates mindfulness practice, improving concentration and stress reduction (Navarro-Haro et al., 2017).

Indeed, the integration of virtual reality into psychological practice is transforming the way we understand and treat mental disorders. Technological innovations have opened new therapeutic and educational possibilities, improving treatment efficacy and research quality. With the continuous advancement of VR technologies, the potential for enhancing clinical practice and understanding cognitive and emotional processes is immense, suggesting a promising future for this innovative therapeutic tool.

8.3   FINAL CONSIDERATIONS AND IMPLICATIONS FOR THE PSYCHOLOGICAL SECTOR

The adoption of VR in psychology promises significant therapeutic, educational, and research benefits, but also raises important ethical and practical issues. These include the accessibility of the technology and the need for specialized training for clinicians who wish to use VR in their interventions. Furthermore, it is crucial to continue evaluating the effectiveness and safety of VR interventions, especially compared to traditional therapies.

The accessibility of VR is a key issue. Despite the decreasing costs of VR devices, the initial investment remains significant, which can be a barrier for healthcare facilities and individuals with limited resources. Specialized training for clinicians is essential to ensure effective use of VR, requiring dedicated training programs and continuous updates on new technologies and methodologies (Riva, 2022).

Despite these challenges, the potential of VR to significantly improve the treatment and understanding of psychological disorders is undeniable. With further research and responsible development, VR could become an essential component of contemporary psychological practice. For example, studies have shown that VR-based therapies can be as effective, if not more effective, than traditional therapies for certain disorders, such as phobias and PTSD (Turner & Casey, 2014).

The increasing adoption of VR in the psychological sector opens new therapeutic perspectives and raises important questions regarding accessibility and effectiveness. While immediate benefits, such as improved mental well-being and reduced cognitive fatigue, are evident, it is essential to consider the long-term implications of such interventions. Continuous research is crucial to establish efficacy standards, especially compared to traditional therapies. Moreover, the issue of accessibility remains relevant, as VR technology requires significant investments in hardware and software development. The availability of portable and cheaper devices could facilitate wider VR dissemination, making it accessible even in resource-limited contexts (Freeman et al., 2017).

Another significant challenge concerns the ecological validity of VR experiences, i.e., the ability of simulations to faithfully reflect real-life situations. Variability in VR protocols can affect therapeutic efficacy and generalizability of results (Vasser & Aru, 2020). It is important to develop standardized protocols to ensure that VR experiences are consistent and reproducible across different studies and clinical applications.

The use of VR also raises important ethical issues regarding the protection of personal data and the management of sensitive information collected during virtual experiences. It is essential to adopt effective data management practices to ensure user security and confidentiality (Madary & Metzinger, 2016). Collecting biometric and behavioral data through VR can offer new research opportunities but also new responsibilities to ensure that such data is protected against misuse.

Finally, the immersiveness of VR can lead to technological addiction phenomena with negative consequences on users' mental and physical health. Promoting responsible VR use is therefore fundamental to preventing such risks (Billieux et al., 2015). Users must be educated about the potential risks of excessive VR use, and clear limits must be established to ensure healthy and balanced use.

The implications of virtual reality for the psychological sector are profound and manifold. VR enables more controlled and repeatable experiments, improving the quality of collected data and the validity of psychological studies. For example, using VR to study problem-solving has revealed how physiological activation can negatively affect cognitive performance, contrary to initial expectations (Marchioro & Benatti, 2022). This demonstrates how VR can offer new insights into cognitive processes that would be difficult to obtain with traditional methods.

VR represents a powerful tool for exposure therapy and other therapeutic interventions, overcoming the logistical and ethical limits of real environments. This not only improves treatment accessibility but also their effectiveness and safety (Botella et al., 2015). The ability to simulate realistic and controlled scenarios allows for personalized and adaptable treatments to the specific needs of each patient, improving therapeutic outcomes.

VR offers new training modalities for psychologists and therapists, allowing them to simulate complex clinical scenarios and develop skills in a safe and controlled environment (De Freitas & Maharg, 2011). VR-based training can include the observation and practice of advanced therapeutic techniques, interaction with avatars simulating patients with various psychological disorders, and receiving immediate and detailed performance feedback. This approach can significantly improve the preparation and confidence of mental health professionals.

In conclusion, it can be stated that virtual reality is transforming the field of psychology, offering new perspectives for research and clinical practice. The technical and methodological innovations introduced by VR promise to revolutionize the way we understand and treat human cognitive and emotional dynamics, opening new avenues for more effective interventions and in-depth studies. The continuous evolution of VR technologies and their integration into psychological contexts represents a promising field of exploration for the future of psychology. However, it is essential to address the ethical and practical challenges associated with the use of VR responsibly, ensuring that the benefits of this technology are accessible and safe for all.

8.4   TAKE HOME MESSAGES

The evidence presented clearly indicates that VR not only improves motivation and engagement but also has profound implications for the psychological sector. Moreover, when combined with psychotherapy, VR's ability to provide personalized experiences is crucial to maximizing the effectiveness of therapeutic interactions (Makransky & Lilleholt, 2018). Additionally, the simulation of natural environments through VR offers an innovative means to promote psychological well-being and combat demotivation, suggesting that this technology could become a fundamental part of future educational and therapeutic practices (Kjellgren & Buhrkall, 2010).

VR's ability to create immersive and interactive experiences is key to engaging users in therapeutic, educational, and restorative activities. For example, some studies mentioned have demonstrated that children are more motivated to participate in therapeutic sessions that include game and interactivity elements, which can lead to significant improvements in their cognitive and social skills (Parsons et al., 2017). Using VR to simulate school and learning environments allows children to explore and learn in a way that is both fun and educational, increasing the effectiveness of teaching strategies.

The integration of VR into psychological practice represents a significant evolution in treating emotional loss and grief. The evidence presented in the fifth chapter indicates that VR can significantly improve the effectiveness of traditional therapies, offering a more personalized and engaging approach. The ability to create immersive therapeutic environments and facilitate emotional regulation makes VR a valuable tool for therapists. The possibility of immersing patients in virtual environments that evoke positive memories or simulate interactions with deceased family members can help process grief more effectively and promote emotional well-being (Riva et al., 2016).

In the future, research should continue to explore the use of VR in combination with other therapeutic modalities, such as cognitive-behavioral therapy with or without exposure, to treat a broader range of mental health issues. Combining VR with biofeedback techniques, for example, could offer new methods to monitor and improve emotional regulation and stress control in patients. With the continuous evolution of VR technology, the potential to enhance therapeutic practice and psychological well-being is immense, suggesting a promising future for this innovative therapeutic tool.

Despite significant progress, using virtual reality in psychology still presents several challenges that must be addressed to fully exploit its potential. The high costs of VR devices and the need for technical training limit its widespread adoption, especially in clinical and academic settings with limited resources (Riva, 2022). However, technological evolution and cost reduction could facilitate broader dissemination in the future. Developments in portable VR devices and open-source software could make this technology more accessible to a wider audience, including therapists and researchers in resource-limited areas.

Another challenge concerns the need to standardize VR intervention protocols to ensure treatments are effective and safe. Variability in VR protocols can affect therapeutic outcomes and the generalizability of studies, making it essential to develop clear and evidence-based guidelines (Vasser & Aru, 2020). Ongoing research is needed to identify best practices and evaluate the long-term effectiveness of VR-based interventions.

With a conscious and responsible approach, these innovations can bring significant benefits to society, improving quality of life and opening new frontiers in psychological research and professional practice. However, it is essential to consider the ethical and social implications of widespread VR use, ensuring that the technology is used safely and beneficially for all. Protecting personal data and managing sensitive information collected during virtual experiences are crucial aspects that require strict regulation to prevent misuse and ensure user privacy (Madary & Metzinger, 2016).

Finally, VR's immersiveness can lead to technological addiction phenomena with negative consequences on users' mental and physical health. Promoting responsible VR use is fundamental to preventing such risks (Billieux et al., 2015). Educating users about the potential risks and establishing clear limits on VR use can help ensure this technology is used healthily and balanced.

In conclusion, virtual reality is transforming the field of psychology, offering new perspectives for research and clinical practice. The technical and methodological innovations introduced by VR promise to revolutionize the way we understand and treat human cognitive and emotional dynamics, opening new avenues for more effective interventions and in-depth studies. The continuous evolution of VR technologies and their integration into psychological contexts represents a promising field of exploration for the future of psychology. With careful and responsible management, VR has the potential to become a fundamental tool for improving mental health and psychological well-being globally.

REFERENCES

Anderson, P. L., Price, M., Edwards, S. M., Obasaju, M. A., Schmertz, S. K., Zimand, E., & Calamaras, M. R. (2017). Virtual reality exposure therapy for social anxiety disorder: A randomized controlled trial. *Journal of Consulting and Clinical Psychology, 85*(3), 238–249. https://doi.org/10.1080/1040041 9.2017.1302741

Argüero-Fonseca, A., Martínez Soto, J., Barrios Payán, F. A., Villaseñor Cabrera, T. D., Reyes Huerta, H. E., González Santos, L., Aguirre Ojeda, D. P., Pérez Pimienta, D., Reynoso González, O. U., & Marchioro, D. (2023). Effects of an n-back task on indicators of perceived cognitive fatigue and fatigability in healthy adults. *Acta Biomed, 94*(6), 1–12. https://doi.org/10.23750/abm. v94i1.15649

Bailenson, J. (2018). Protecting nonverbal data tracked in virtual reality. *Nature Human Behaviour, 2*(7), 431–432. https://doi.org/10.1038/ s41562-018-0400-5

Billieux, J., Thorens, G., Khazaal, Y., Zullino, D., Achab, S., & Van der Linden, M. (2015). Problematic involvement in online games: A cluster analytic approach. *Computers in Human Behavior, 43*, 242–250. https://doi. org/10.1016/j.chb.2014.10.055

Botella, C., Serrano, B., Baños, R. M., & Garcia-Palacios, A. (2015). Virtual reality exposure-based therapy for the treatment of post-traumatic stress disorder: A review of its efficacy, the adequacy of the treatment protocol, and its acceptability. *Neuropsychiatric Disease and Treatment, 11*, 2533–2545. https://doi. org/10.2147/NDT.S89542

Botella, C., Serrano, B., Baños, R. M., & García-Palacios, A. (2017). Virtual reality exposure-based therapy for the treatment of post-traumatic stress disorder: A review of its efficacy, the adequacy of the treatment protocol, and its acceptability. *Neuropsychiatric Disease and Treatment, 13*, 2533–2545. https://doi. org/10.2147/NDT.S118620

Chen, J., Fu, Z., Liu, H., & Wang, J. (2024). Effectiveness of virtual Realityon learning engagement: A meta-analysis. *International Journal of Web-Based Learning and Teaching Technologies, 9*(1), 1–14. https://doi.org/10.4018/IJWLTT.334849

Culbertson, H., Schorr, S. B., & Okamura, A. M. (2018). Haptics: The present and future of artificial touch sensation. *Annual Review of Control, Robotics, and Autonomous Systems, 1*(1), 385–409. https://doi.org/10.1146/annurev-control-060117-105043

De Freitas, S., & Maharg, P. (2011). *Digital games and learning*. Continuum International.

Emmelkamp, P. M., Krijn, M., Hulsbosch, A. M., de Vries, S., Schuemie, M. J., & van der Mast, C. A. (2002). Virtual reality treatment versus exposure in vivo: A comparative evaluation in acrophobia. *Behaviour Research and Therapy, 40*(5), 509–516. https://doi.org/10.1016/s0005-7967(01)00023-7

Freeman, D., Reeve, S., Robinson, A., Ehlers, A., Clark, D., Spanlang, B., & Slater, M. (2017). Virtual reality in the assessment, understanding, and treatment of mental health disorders. *Psychological Medicine, 47*(14), 2393–2400. https://doi.org/10.1016/S2215-0366(17)302998

Freina, L., & Ott, M. (2015). A literature review on immersive virtual reality in education: State of the art and perspectives. *Conference Proceedings of eLearning and Software for Education (eLSE)*. https://doi.org/10.12753/2066-026X-15-020.

Garcia-Ruiz, M. A., Edwards, A., El-Seoud, S. A., & Aquino-Santos, R. (2008). Collaborating and learning a second language in a wireless virtual reality environment. *International Journal of Mobile Learning and Organization, 2*(4), 369–377. https://doi.org/10.1504/IJMLO.2008.020689

Garrett, B., Taverner, T., & McDade, P. (2017). Virtual reality as an adjunct home therapy in chronic pain management: An exploratory study. *JMIR Medical Informatics, 5*(2), e11. https://doi.org/10.2196/medinform.7271

Jordan, J. R., & Neimeyer, R. A. (2003). Does grief counseling work? *Death Studies, 27*(9), 765–786. https://doi.org/10.1080/07481180390229203

Kjellgren, A., & Buhrkall, H. (2010). A comparison of the restorative effect of a natural environment with that of a simulated natural environment. *Journal of Environmental Psychology, 30*(4), 464–472. https://doi.org/10.1016/j.jenvp.2010.01.011

Lau, W. K., Chiu, T., Ho, K., Lo, S., & Luk, M. (2017). The effectiveness of virtual reality-based training on functional outcomes for children with cerebral palsy: A systematic review and meta-analysis. *Developmental Medicine & Child Neurology, 59*(7), 717–725. https://doi.org/10.1111/dmcn.13419

Madary, M., & Metzinger, T. K. (2016). Recommendations for good scientific practice and the consumers of VR-technology. *Frontiers in Robotics and AI, 3*. https://doi.org/10.3389/frobt.2016.00003

Makransky, G., & Lilleholt, L. (2018). A structural equation modeling investigation of the emotional value of immersive virtual reality in education. *Educational Psychology Review, 30*(3), 873–889. https://doi.org/10.1037/edu0000241

Maples-Keller, J. L., Bunnell, B. E., Kim, S.-J., & Rothbaum, B. O. (2017). The use of virtual reality Technology in the Treatment of anxiety and other psychiatric disorders. *Harvard Review of Psychiatry, 25*(3), 103–113. https://doi.org/10.1097/HRP.0000000000000138

Marchioro, D. M., Argüero-Fonseca, A., Bounous, M., & Benatti, F. (2023). Effectiveness of images with high-potential restorative in Virtual Reality to reduce acute cognitive fatigue in undergraduate students. In *18th European congress of psychology - ECP. Psychology: Uniting communities for a sustainable world* (p. 274). European Federation of Psychologists' Associations.

Marchioro, G., & Benatti, F. (2022). *Processes of planning in a virtual reality experience: Link between arousal and problem solving.* Poster presented at the *17th European Congress of Psychology (ECP) "Psychology as the Hub Science: Opportunities and Responsibility"—Lubiana (Slovenia).*

Maskey, M., Lowry, J., Rodgers, J., McConachie, H., & Parr, J. R. (2014). Reducing anxiety in children with autism spectrum disorder using a virtual reality environment: A pilot study. *PLoS One, 9*(7), e100374. https://doi.org/10.1371/journal.pone.0100374

Moreno, A., Wall, K. J., Thangavelu, K., Craven, L., Ward, E., & Dissanayaka, N. N. (2019). A systematic review of the use of virtual reality and its effects on cognition in individuals with neurocognitive disorders. *Alzheimer's & Dementia: Translational Research & Clinical Interventions, 5,* 834–850. https://doi.org/10.1016/j.trci.2019.09.016

Moyle, W., Jones, C., Dwan, T., & Petrovich, T. (2018). Effectiveness of a virtual reality Forest on people with dementia: A mixed methods pilot study. *The Gerontologist, 58*(3), 478–487. https://doi.org/10.1093/geront/gnw270

Navarro-Haro, M. V., López-del-Hoyo, Y., Campos, D., Linehan, M. M., Hoffman, H. G., et al. (2017). Meditation experts try virtual reality mindfulness: A pilot study evaluation of the feasibility and acceptability of virtual reality to facilitate mindfulness practice in people attending a mindfulness conference. *PLoS One, 12*(11), e0187777. https://doi.org/10.1371/journal.pone.0187777

Neimeyer, R. A. (2001). Meaning reconstruction & the experience of loss. *American Psychological Association.* https://doi.org/10.1037/10397-000

Parsons, S., & Cobb, S. (2011). State-of-the-art of virtual reality technologies for children on the autism spectrum. *European Journal of Special Needs Education, 26*(3), 355–366. https://doi.org/10.1007/s10639-011-9181-8

Parsons, T. D. (2015). Virtual reality for enhanced ecological validity and experimental control in the clinical, affective and social neurosciences. *Frontiers in Human Neuroscience, 9,* 660. https://doi.org/10.3389/fnhum.2015.00660

Parsons, T. D., Riva, G., Parsons, S., Mantovani, F., Newbutt, N., Lin, L., Venturini, E., & Hall, T. (2017). Virtual reality in pediatric psychology. *Pediatrics, 140*(Suppl 2), S86–S91. https://doi.org/10.1542/peds.2016-1758I

Riva, G. (2022). Virtual reality in clinical psychology. *Comprehensive Clinical Psychology*, 91–105. https://doi.org/10.1016/B978-0-12-818697-8.00006-6

Riva, G., Baños, R. M., Botella, C., Mantovani, F., & Gaggioli, A. (2016). Transforming experience: The potential of augmented reality and virtual reality for enhancing personal and clinical change. *Frontiers in Psychiatry, 7*, 164. https://doi.org/10.3389/fpsyg.2016.01266

Riva, G., Wiederhold, B. K., & Mantovani, F. (2019). Neuroscience of virtual reality: From virtual exposure to embodied medicine. *Cyberpsychology, Behavior, and Social Networking, 22*(1), 82–96. https://doi.org/10.1089/cyber.2017.29099.gri

Rothbaum, B. O., Hodges, L. F., Kooper, R., Opdyke, D., Williford, J. S., & North, M. M. (1995). Effectiveness of computer-generated (virtual reality) graded exposure in the treatment of acrophobia. *American Journal of Psychiatry, 152*(4), 626–628. https://doi.org/10.1176/ajp.152.4.626

Slater, M. (2009). Place illusion and plausibility can lead to realistic behaviour in immersive virtual environments. *Philosophical Transactions of the Royal Society of London. Series B, Biological Sciences, 12*(364(1535)), 3549–3557. https://doi.org/10.1098/rstb.2009.0138

Turner, W. A., & Casey, L. M. (2014). Outcomes associated with virtual reality in psychological interventions: Where are we now? *Clinical Psychology Review, 34*(8), 634–644. https://doi.org/10.1016/j.cpr.2014.10.003

Vasser, M., & Aru, J. (2020). Guidelines for immersive virtual reality in psychological research. *Current Opinion in Psychology, 36*, 71–76. https://doi.org/10.1016/j.copsyc.2020.04.010

INDEX

GPSR Compliance
The European Union's (EU) General Product Safety Regulation (GPSR) is a set
of rules that requires consumer products to be safe and our obligations to
ensure this.

If you have any concerns about our products, you can contact us on

ProductSafety@springernature.com

In case Publisher is established outside the EU, the EU authorized
representative is:

Springer Nature Customer Service Center GmbH
Europaplatz 3
69115 Heidelberg, Germany